A Little Land and a Living

by Bolton Hall

with an introduction by Roger Chambers

This work contains material that was originally published in 1908.

This publication was created and published for the public benefit, utilizing public funding and is within the Public Domain.

This edition is reprinted for educational purposes and in accordance with all applicable Federal Laws.

Introduction Copyright 2017 by Roger Chambers

Self Reliance Books

Get more historic titles on animal and stock breeding, gardening and old fashioned skills by visiting us at:

http://selfreliancebooks.blogspot.com/

Introduction

I am pleased to present yet another title on Gardening.

The work is in the Public Domain and is re-printed here in accordance with Federal Laws.

As with all reprinted books of this age that are intended to perfectly reproduce the original edition, considerable pains and effort had to be undertaken to correct fading and sometimes outright damage to existing proofs of this title. At times, this task is quite monumental, requiring an almost total "rebuilding" of some pages from digital proofs of multiple copies. Despite this, imperfections still sometimes exist in the final proof and may detract from the visual appearance of the text.

I hope you enjoy reading this book as much as I enjoyed making it available to readers again.

Roger Chambers

Melepepo Indicus flauus
cum semine suo aperto
ubi conspiciuntur semina.

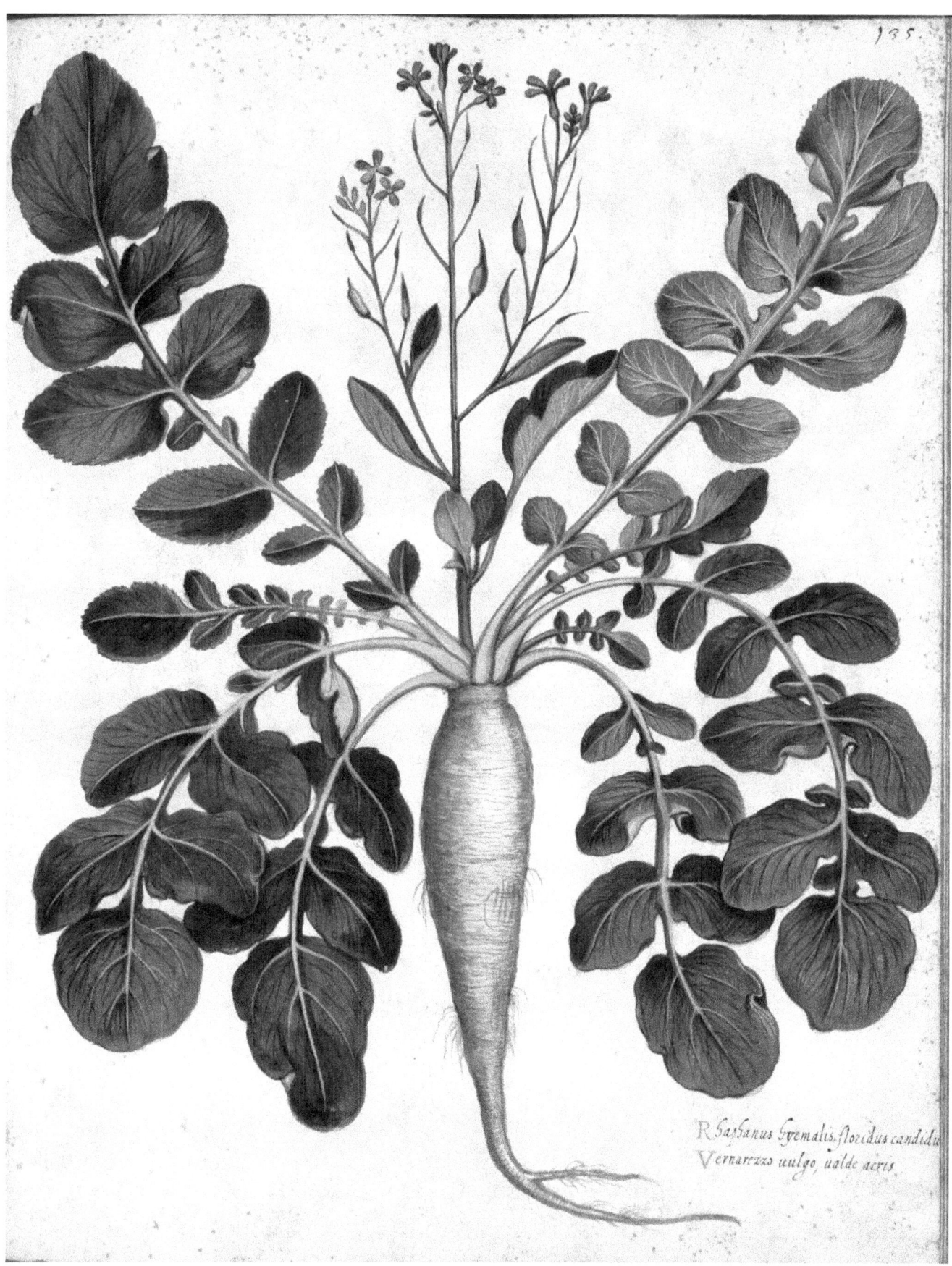

R. Raphanus hyemalis, floridus candidus
Vernarezzo vulgo, valde acris

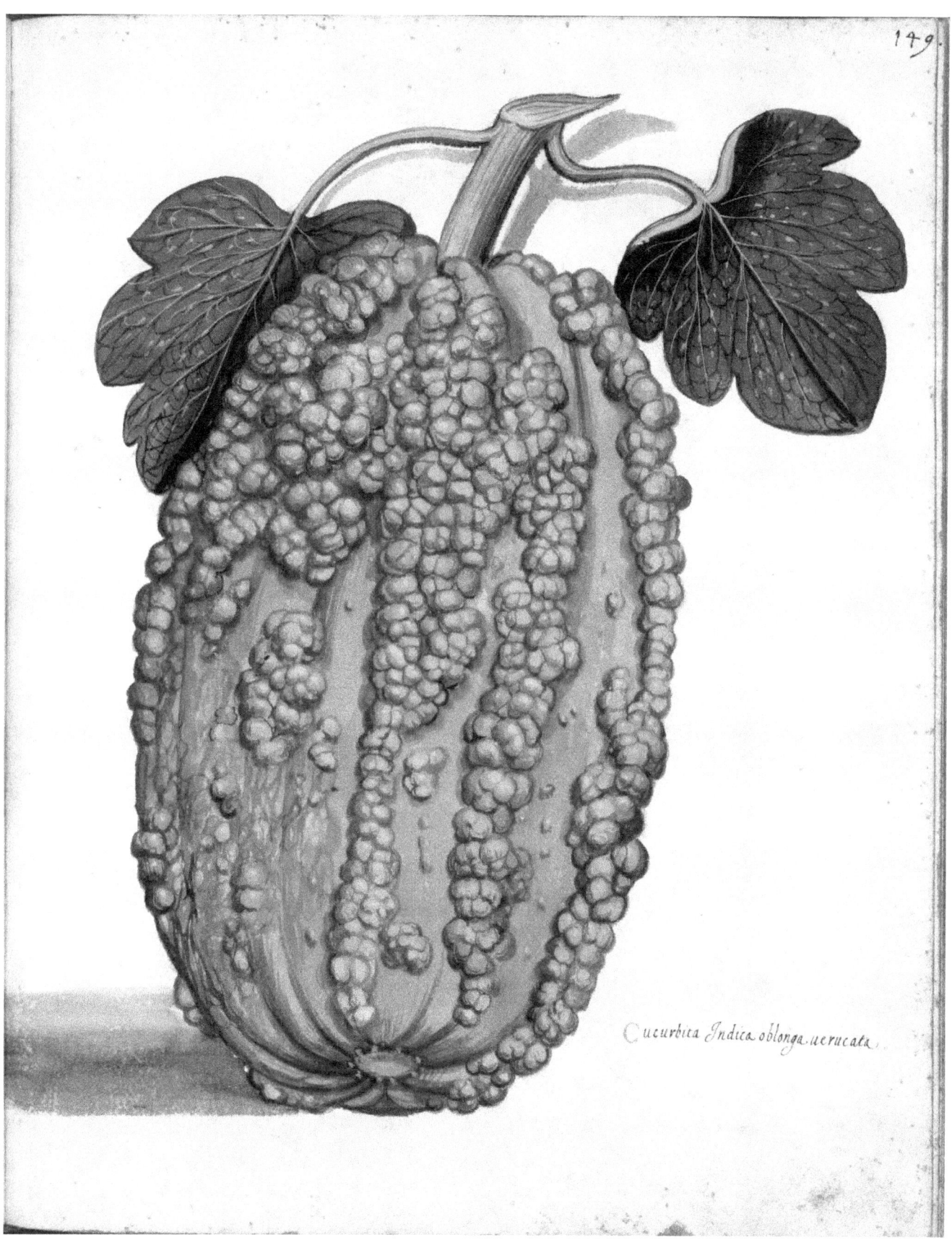
Cucurbita Indica oblonga verucata

Cucurbita maior nostras urbana, colore citrino.

FOREWORD

INTEREST in the "little lands" from which men may make a living continues to grow and spread. A money panic does more than scare people—it sets them thinking how they can protect themselves against a recurrence of this thing. That necessarily turns their thoughts to the land as the source of wealth and independence. It is because of this growing desire on the part of the people to know what can be done with small areas, that the author has written this book. Every chapter has been submitted to some expert for correction and revision, and the author gladly acknowledges indebtedness to Mr. George T. Powell, President of the Agricultural Experts Association; Professor W. G. Johnson, of the Orange Judd Co.; Mr. R. F. Powell, Superintendent of the Philadelphia Vacant Lots Association; Miss Kate Sanborn, Mr. Howard

Goldsmith, of the Suffolk Farms Co.; Mr. Samuel Milliken, and others for valuable aid and suggestions. The footnotes give credit to other sources of information.

The reception accorded by press and public to my book, *Three Acres and Liberty,* which Macmillan published a year ago, was a pleasing proof of the interest already awakened in this matter. Six editions of that book have been issued, and indications are that others will follow.

But no one volume could begin to exhaust so fruitful a subject, and the readers of *Three Acres and Liberty* will not find *A Little Land and a Living* in any sense a repetition of its predecessor. The reasons for its publication at this time are numerous and cogent, many of them being set forth in Mr. Borsodi's letter, which follows this foreword. Others may wisely be left to its readers to infer.

Those who are facing the problem of rearing a family on a weekly wage, with the purchasing power of the dollar decreasing, will find much in this book to encourage them to reach out for a better, saner living, through cultivating the little lands.

Those who know most of farming believe that it is only a question of once learning what to do and how to do it, to draw many of the city workers to the outlying lands. This *A Little Land and a Living* aims to do; not to induce the unfamiliar to rush headlong into farming, but to encourage those who feel the pressure of city life to study how they may get away from the overcrowded city into nearby country, where the gardens may first be made an adjunct to the income and later, perhaps, prove the source of the income.

Mr. George T. Powell writes:

"You have brought together many facts and information that should be helpful and be an aid to many who, for the want of specific information, do not realize what they might do on a small land-holding.

"If there could be lectures given on this subject in the tenement districts it would be of special value. I advocated this in a lecture at the United Charities Building on 'How to Help the City Poor to Get Out to the Land.'

"They do need specific instruction, first, where

FOREWORD

they are, and should then be given some aid in reaching cheap land, so that they might make a start. I believe that if a regular course of instruction of the simple, plain kind could be given in these parts of the city it would be of real value."

That is sound sense.

BOLTON HALL.

CONTENTS

"A LITTLE LAND AND A LIVING" . . . . . *Frontispiece*

                                                                             PAGE

THE LETTER THAT PROMPTED THIS BOOK WRITTEN TO THE AUTHOR BY WILLIAM BORSODI . 17

    A Many-Sided Problem of International Scope—How Poverty, Insanity, Vice, Might Have Been Prevented—Objections to the Cry "Back to the Farm"—You Show How to Make Farm Life Pay—Does the City Give Comfort?—The Curse of "Credit"—Where "Home" is Not "Sweet Home"—No "Sweetness and Light" Here—"Oh, the Cold and Cruel Winter"—Waiting for Jobs That Do Not Come—"The Bread Line" and Other Lines—The Misery Seen in Missions—"Three Acres" Would Mean "Liberty"—City Amusements and Other "Elevating" Things—Statistics That Give False Impressions—The Improvidency of the Poor—Dire Necessity—"The Great White Way" as Seen by Rich and Poor—The Dance Halls and the Street—No Real Suffering and Destitution on Farms—The Educated and Inexperienced Adapted for Agriculture—Scope for "Knowledge Learned of Schools"—Toil and Payment of Farm Life—The World's Ideal Farming Country—Agriculture's Golden Age—Independence in Agriculture—Land Enough for a Living—One Point of Honest Difference—Who Should Go Back to the Farms?—Who Would Benefit by Such a Movement?—Semi-Agricultural Colonies—Other Letters from Railroad Men—A Pleasing Tendency—What I have Done—Write a New Book—The Cause and the Remedy—Foreigners Live More Cheaply—Concerted Action Necessary—Many Men of Many Minds—The Real Hope.

CHAPTER I

LIFE, NOT MERELY MAKING A LIVING 77

    Opportunities Past and Present—Chances Lost—New Chance—Intelligence—Slums a Symptom—The Origin of Wealth—Capital—An Acre Enough—Possibilities—Thoroughness—Available Land—New Methods—Intensive Farming—Trucking—Dairying—How They Do in Japan—Denmark—Jersey—We Can Do Better.

CONTENTS

CHAPTER II

BUYING A GARDEN 85

>Ownership — Sources of Information — Agents — Where Not to Go—Fertility Secondary—Soil Never Barren—What to Buy—Run-down Farms—Cheapness Relative—Where to Buy—How to Buy—Don't Fear Debt—Borrowing Money—Insurance—The Monopoly—Unproductive Lots—Use and "Improvements"—Make Unused Land Pay—To Buy or to Build.

CHAPTER III

VACANT LOT GARDENING 95

>Charity vs. Self-Help—Opportunities—Assistance—Co-operation—Little Government Needed—Cost—Health and Success—A Variety of Returns—One-half Acre — Large Profits — Land, not Capital, Neecessary.

CHAPTER IV

REASONABLE PROSPECTS 105

>Living Costly—Hunger—Value of Food Products—What a Man Wants to Know—Fortune in an Acre—Stony Wold Record—Old Methods—New Methods—Acre Profits — Irrigated — Shearer's Success — O'Brien's—Hartman's—Small Gardens—A Woman's Patch—A 40 x 50 Garden—A City Backyard—Five Cents Per Square Foot—Glade Lands—An Illinois Plot—A Michigan Experiment—Farmers as Robbers — Youthful Gardners — With Brains — Average Yields—Census Reports—What Averages Imply—Bailey's Estimate—Philadelphia Gardners—Uncommon Vegetables—Other Callings Similar—Scientific Farming—The Farmer's Returns—What Could be Made—What He Makes.

CHAPTER V

RECORD YIELDS 131

Possibilities—Production and Cost—A Standard—Records—A Garden for Five—School Gardens—Small Plots—Yields of "Poor" Soil—Celery—Texas Onions—Corn Record—Strawberry Yields—Rhubarb—Christmas Trade—A House-cellar Patch—Forcing Cellar—Onions by New Culture—Asparagus.

CHAPTER VI

WAYS OF WORKING 149

The Home Garden—First—Working the Soil—A Beginner's Garden—Save Labor—Fruit—Planting and Transplanting—One Crop Risky—Companion Cropping—Specialties—Marketing—Succession Crops—Deepening Cultivation—Subsoiling—Soil Enriches Itself—Insects Smothered by Ashes—Winter Plowing to Kill—Spraying—Fertilizers—Irrigation—Drainage—Dry Farming—Fancy Packing.

CHAPTER VII

MONEY AND TIME REQUIRED 173

The Teacher of Fools—Succeess—Failure—"Think-box" Secrets—Large Capital Not Needed—Specialties as Money-makers—The Value of Money—Equipment—Outlay and Income—Five-acre Investment—Ten Acres Costly—Cost of Starting—Union Wages in Different Cities—Hand vs. Horse Cultivation—Crops Every Month—The Condition of the Farmer.

CHAPTER VIII

GROWING UNDER GLASS 187

Early Vegetables—A Sample Hot-bed—Heating by Fire—Cost and Returns—Flowers Better Than Vegetables—Greenhouses—Success.

CONTENTS

CHAPTER IX

PAGE

ANIMALS FOR PROFIT 194

Animals on the Farm—A Snail Park—Frogs—Turtles—Bass—Pheasants—Dogs—Cats—Silver Foxes—Expenses and Receipts—The Busy Bee.

CHAPTER X

FRUIT GROWING 211

American Supremacy — Development — Improvement—Apples, Quality, Thinning—Peaches—Chances of Success—Protecting Grapes—Pears—Plums—Quinces—Cherries—Persimmons—Small Fruit—Extra Culture—Strawberries—Bush Fruit—Exceptional Returns.

CHAPTER XI

HORTICULTURE 227

The Market—Violets—Greenhouses—Diseases—Roses—Chrysanthemums—Poppies—Flowers in the Street—Sweet Peas and Wild Flowers—Orchids—Plants for Renting—Floriculture.

CHAPTER XII

BUILDING 237

Clearing the Land—Lumbermen Buy—Profitable Trees—Building a Home—"Hickory Bungalow"—Portable Dwellings—Remodeled Buildings—Comfortable Cabins—Water Supply.

CHAPTER XIII

CO-OPERATION IN OPERATION 247

Toil Without Reward—"Back to the Land"—How to Get Land—Co-operation—Man's Natural Job—Organization—In Europe—In America—Fellowship Farm—The Arden Colony—Farming a Business—Changes Imminent.

CHAPTER XIV

TO START SANITARIUM WORK 259

Outdoor Life Effective—Dr. Trudeau's Plan—Bad Conditions—Convalescents Work—The Earth for Men—Location Important—The Superintendent's Value—The Money Needed—Supplementary Industries—Preserving, Baking, Selling, etc.—Start Now.

CHAPTER XV

THE PROFESSION OF FARMING 269

Agriculture, Past and Future—The Average Farmer—The Boy and the Garden—The Trained Farmer—Salaries Await Him—Fresh Discoveries—Fields for Investigation—Grown from the Best—The Profit of the Earth—Monopoly Conditions—The Value of the Farm—How to Proceed—The Aim of This Book.

ABOUT ANOTHER BOOK ON "BACK TO THE LAND" . 289

My Dear Mr. Hall:

In making myself acquainted with your book, " Three Acres and Liberty," I have done a great deal more than just fingering it. I am convinced that this work will fill a pronounced want and become, in its line, an epoch-maker.

I thank you for referring to me in your autograph introduction as your " co-worker." You certainly hit the nail on the head. So far as concerns the desire of getting the people back to the farm, I am your " forerunner," and nothing would make me happier than to be your "co-worker" in a practical way and to see crowds taking advantage of the proposition your book so ably puts before them. If I were wealthy I would circulate this book broadcast. The needy would profit by its suggestions, and I should feel amply repaid for the money expended.

More than thirty-five years ago, I started to think and to reflect. Since then, naturally, I have found out my early mistakes. I was attracted for the most part of that long period by the teachings of our conservative, reactionary, liberal Henry George, and even " my " own sacred " policies " and ideas attracted me. Socialism in Germany is very good; anarchism in Russia is not a bit less good, but these " isms " are not good everywhere, nor are they good for all the time. There are other remedies and policies for the alleviation of the degrading social conditions that confront us everywhere; but there is one which is as efficient now as ever, in every land, in all places and at all times—farm life.

A MANY-SIDED PROBLEM OF INTERNATIONAL SCOPE.

I have in mind a problem, complicated and far-reaching in its being and influence, because it vitally concerns every large city of all countries; a problem of international scope—the over-crowding of large cities with unadaptable laborers, all seeking a livelihood where the opportunities don't increase in proportion. And as soon as practical methods expedite the solution of this many-sided problem, the lives and destinies of millions of working men, women and children now living only half-lives will be enlarged and improved.

HOW POVERTY, INSANITY, VICE, MIGHT HAVE BEEN PREVENTED.

In all countries and climates I have found that all people would benefit to a surprising degree by keeping close to the land, and that thousands of those now in the hospitals, lunatic asylums and penitentiaries, or who are living in abject poverty, or on the borders of that state, could have kept in healthful conditions had they gone back to the land in good time and remained there.

OBJECTIONS TO THE CRY "BACK TO THE FARM."

Mine is not a mere judgment or opinion based on a moment's observation or consideration. I have spoken with all classes and grades—the "under dogs," the middlemen, professional men, statesmen, philanthropists and sociologists, socialists and retired farmers, and even with politicians, only to find that the majority of them have something against the "back to the farm" proposition. The preponderance of opinion is that while it is desirable to own a plot of land and a home, farming does not pay, and that you cannot keep the people on the farm once their "ambitions" are aroused by the alluring so-called "opportunities" of the city, and that you cannot induce them to go back to the farm again after a taste of city life, however disagreeable.

YOU SHOW HOW TO MAKE FARM LIFE PAY.

Your book proves that farm life does pay. If the city dweller can be made to see this and to realize what a small percentage city life pays, it will be easy to arouse sufficient interest in a movement which will eventually bring us to healthful conditions and to a race that will feel content on the farm. Money-making attracts every one; those with much money, yet want more; those with some, wish to increase it; and those with little or none are ever in a maelstrom of passion for money. Certainly it would attract a large number of those who now shovel snow for the mere purpose of self-preservation, and those of starved physiques, who can only look on while the world moves, and others whose life is incessant toil with practically nothing to show for it but ill health, a wife and children in stuffy and gloomy half-furnished, broken down boxes of flats in "houses" where twenty to forty families live in similar circumstances.

DOES THE CITY GIVE COMFORT?

New York, Chicago, Philadelphia, and practically every large city in the United States, are the scenes of heart-rending struggles for existence. Not only in our country, but in every country is this so. It is only at our own doors that such scenes strike our sympathies and awaken our pity, and bring to our face a glow of shame that such is the living of millions of our countrymen.

Take a trip through the "living" sections of our large cities—not only the slums and sections where foreigners herd together like so many cattle. I mean the homes of the poor American bred or the naturalized Americans, be they Irish, Scotch, English, German, Italian, or what not—those who are making the United States what it is; the toilers, proud to call themselves Americans. Yes, take a trip down any of those avenues or through those streets—they are found everywhere. Look into the apartments through the dilapidated shutters, you will see all that is necessary. You go by hundreds of houses not properly protected by the windows against cold, nor against the burning sun in summer. The landlord gets "only" enough to pay his taxes, to make only absolutely necessary repairs on the buildings—more to protect the property from dilapidation than to give comfort to the tenants—and for a few incidentals, so he cannot afford to keep them in good shape. In most cases you will be shocked by the scanty and broken-down furnishings that fill these rooms; or, if the furnishings are in any way respectable-looking, you are safe in believing them to be the outfit of some instalment house, on whose basis many of the tenants of three or four rooms furnish their homes.

THE CURSE OF "CREDIT"

There are hundreds of these house-furnishing firms selling on credit in New York and other cities. Some of these houses carry thousands of such accounts, so you can judge for yourself how many persons are sleeping on what the marshal may take away in the morning because four weeks' instalments of fifty cents or a dollar are in arrears.

WHERE "HOME" IS NOT "SWEET HOME."

During the summer their houses are practically untenantable. In the evenings you will find the occupants sitting in the streets, on the roofs, in the parks, or in other places more attractive to the body and soul. "There's no place like home" is a farce to more than 75 per cent. of the inhabitants of every large city.

Has it not come to a bad pass when cities like New York or Chicago and others have to grant special permission for the millions of tenement dwellers to utilize the public parks and piers and playgrounds as lodging houses, because their homes are so stifling during the dog days as to make it a torture for a human being to sleep in them? Is it not as revolting to see families sleeping on fire-escapes and roofs in summer; or in winter to read of persons freezing to death in the streets and even in their scanty, ill-protected houses? Or to listen to the thousand appeals from as many charitable organizations and institutions, calling for aid in the relief of the untold misery of the poor, all the year round?

NO "SWEETNESS AND LIGHT" HERE.

Picture the "happy homes" of hundreds of thousands of poor, who live in dilapidated structures that occupy the space needed for ventilation and light—the back yards. Thus sandwiched in they are surrounded on all sides by higher houses, shut off from sunlight all day, their only view of the outside world being the back windows of the flats of families little better situated than themselves; their avenue of entrance into their little world being a dark, gloomy, foul alley, often used as a refuse dump. There is where you can also see hard-working Americans "living"—freezing well-nigh to death in winter and fairly roasting alive in summer! And these are the people who are the first objects of charity if the man of the house be out of work. They live without thought of the morrow; they cannot do otherwise, for their income hardly covers living expenses. There is where the children, if they survive, are compelled to help out by working in factories at an early age, stunting their youth, undermining their health, and befitting themselves for only the meanest sort of labor in after years.

"OH, THE COLD AND CRUEL WINTER!"

This winter proved to be one of the severest in misery for the poor and unemployed—every winter is, more or less. But this year's scenes of misery were more pronounced. Everywhere workers were cast out, the army of street beggars and vagrants and fakirs increased apparently an hundredfold, and the free municipal and mission lodging houses, not counting the five, ten and fifteen cent varieties, turned away thousands of men and women, whose only refuge from the cold and storms were alleys, doorways, vacant lots and unused trucks. In Pittsburg a conservative observer estimated the unemployed alone at 15,000; in Chicago 75,000, while here in New York it was reported that 90,000 members of organized labor were without jobs and that there were 30,000 vagrants.

WAITING FOR JOBS THAT DO NOT COME.

At the time the mercury was at its lowest, I happened to pass along Union Square early in the morning, plowing my way through several inches of snow, which had fallen the night before. I noticed on one corner a group of four poorly-clad men, hands stuffed deep into pockets, while they shivered and shivered, nose, cheeks and lips blue and purple from the cold. From what I could see they were trying to secure work at shoveling snow. Three hours later, returning down-town, I passed the same group—still waiting for the " job."

"THE BREAD LINE" AND OTHER LINES.

The old bread line at midnight in New York is a favorite topic of the writer on sociological subjects, but such strings of workless and starved beings were to be seen everywhere. I saw a new kind of a line after the big snowstorm. I glanced along it and saw that it extended fully a block. Every man in that line was waiting to get the few dollars, or few cents, he was entitled to for his share in cleaning away the snow. Such frost-bitten, hungry looking individuals it has not been my lot to look upon in years.

THE MISERY SEEN IN MISSIONS.

Still another picture of the "abounding comforts" of the big city was to be seen here in the Metropolis, and photographs of it have been scattered far and wide by an enterprising photographer, as if the scene was one to be proud of! The municipal lodging houses being filled, a Bowery mission opened its doors to 500 unfortunates. There were no beds, but these people were permitted to sit in the mission all night until seven a. m., so as to be out of the cold. Many of these found the welcome opportunity to snatch a few hours of sleep in a sitting posture.

A LITTLE LAND AND A LIVING

COLD WAVES, WARM HEARTS.

They should go together at the thought of thousands of women and children in our city, shivering, hungry, sick, through no fault of their own.

They are sent to us by teachers, doctors, churches, city officials, neighbors.

An instance: "I am an ironworker laid off two months ago. My wife is sick; there has not been a crumb in our house for two days. I can't bear to go home."

We relieve suffering at once and then try to get people on their feet.

$20,000 are urgently needed for food, coal, rent, clothes, bedding, medicine. We have over 2,000 families in charge to-day.

N. Y. Association for Improving the Condition of the Poor.

LEND A HAND.

This storm means suffering which will last for days.

You would cheerfully help to get a fallen horse upon his feet. Will you do as much for a man?

Many in our city have heavy burdens and a very insecure footing. An accident, temporary loss of work, brings quick suffering to dependent women and children. We know over 2,500 such families to-day. Will you help relieve at least one; details if desired? $20,000 urgently needed. Send $1.00, $5.00, $10.00, $50.00, $100.00, and let us tell you what it does, to R. S. Minturn, Treas., Room 210, No. 105 East 22d St.

N. Y. Association for Improving the Condition of the Poor

These two advertisements, which have recently been appearing in the New York papers, are appealing on behalf of people who would not be a burden on the public at large—who would not lose their own self-respect, if they would remain on or return to the land.

"THREE ACRES" WOULD MEAN "LIBERTY."

Just why all these people should not live on your "three-acre" farm, even in a large dry goods box rather than increase the value of real estate in Manhattan or Philadelphia or the great city on Lake Michigan, or than swell the ranks of the objects of charity, I cannot see. Why should not these people live upon a farm near the city, to come daily to their business, whatever it be, and occasionally for amusement, and to spend their spare time in the profitable and healthy occupation—farming? At this time when the news of the day gives us startling instances of top-heaviness in population in reports of the destitution almost everywhere, the benefits of the "back to the farm" movement are apparent on its face. If these people for whom the land is crying could be induced to locate in the rural districts, the cause and results of much of this suffering would be done away with. Not only would the country districts be benefited by such a change, but the cities as well. Much of the distress which now prevails in the cities is a detriment to all in these days when country and city are so knit together in a subtle bond of action and interaction.

CITY AMUSEMENTS AND OTHER "ELEVATING" THINGS.

Talk to the young men or women here in New York (of course not ALL of them), especially those born and brought up in town, and before anything else they will tell of the great fun they are having. They work hard, and have long hours. Yet these young people, inclined toward fun, spend every cent they can spare on clothes and amusements.

STATISTICS THAT GIVE FALSE IMPRESSIONS.

I will cite here a few instances which show what the city offers the poor. I call "poor" every one who cannot afford to be sick for a few weeks, or who has not over four weeks' wages in the savings bank. The savings bank reports show how many hundreds of thousands of people have deposited in the hundred odd savings institutions on Manhattan Island alone. If you walk into any worker's home, even on Third

Avenue, you will at once know the real truth of those savings statistics.

While the habit of thrift is being cultivated to a larger extent than heretofore, these statistics leave false impressions; and it is only when you get in touch with the people that go to make up our large cities that you become informed as to the impossibility of saving among workers. There are few bank accounts to be found among the toilers, the clerks, the stenographers, the bookkeepers, the tradesmen, and even some of the professionals of our great cities, who earn barely enough to give them a living, much less a decent home, or to give them opportunities for saving. The younger ones know not the value of money; the call of the great white ways is too strong to resist. They will stint themselves, week in and week out, in order to scrape together enough of their meagre earnings to go to the theatre, dances and dinners. I discussed this subject recently with a young man, nearly twenty years of age, and he told me in all confidence that one day's outing with his "lady friend" a few days previous cost him all told a little over ten dollars—and his salary is just twelve dollars per week! If this was a thing to be indulged in once or twice a year I should be able to see how he does it, but to repeat the same performance about once a month and sometimes more, and all on that small salary, is past comprehension or belief.

THE IMPROVIDENCY OF THE POOR.

When a slack season comes and these young people are thrown out of work they have nothing to fall back on. The task of tiding them over for a few weeks, or until they secure another position, rests upon their family or friends. If not that, then the men go to swell the tide of beggars, paupers and criminals, and the women turn to other forms of social viperism, such as prostitution, where they can " get " more money and better clothes than is afforded through legitimate and healthy occupations. How many a young woman, especially among the clothes workers who earn mean pittances by piece work, has to resort to this means of a livelihood in order to make ends meet, not only for herself, but for an aged mother, possibly, whom circumstances have thrown upon her care! Yet what a common story it is to us living in the big cities.

DIRE NECESSITY.

Passing through Fifteenth Street in New York recently, I noticed a woman and a young girl with small pieces of iron raking the refuse out of ash-cans and putting it into a bag.

It is a revolting scene you have often seen, but you probably turned away, praying for power enough to place these people in a position where such acts would not become a necessity. The woman was poorly dressed, but from her energy and activity she had nothing of the beggar about her. The child, not more than ten years of age, had inflamed eyes, due most likely to the unhealthy occupation. The husband probably was out of work at the time, or was struggling along on five or six dollars a week. From this at least two dollars a week must be paid for rent, so, in order to help along, the wife must do some such work as this in addition to her home duties.

"THE GREAT WHITE WAY" AS SEEN BY RICH AND POOR.

In one of my rambles up Broadway I passed the Metropolitan Opera House just at the time the carriages were discharging ladies exquisitely gowned and resplendent with diamonds and jewels, and gentlemen in evening dress. They, in turn, were being ushered into the lobby, attended by ushers and policemen, while, at the same time, groups of men, women and children, attracted more by a curiosity born of envy than mere idleness, were ordered to move on. Some passed on, and, as though they, in their poverty, must have the same kind of entertainment, entered one of the slot-machine parlors which furnished amusement for as many pennies as their patrons can or cannot afford to spend, and whose pictures, for the most part, are not fit for young people to look at. These people appeared to be families of workingmen who were, in their way, seeing the great white way, and these slot-machine parlors were their avenues of entrance.

THE DANCE HALLS AND THE STREET.

Then again think of the hundreds and hundreds of cheap dance halls—an evil which the papers are loudly decrying—where the children of these poor workers go of an evening to seek amusement, but in reality to be thrown into dens of vice and iniquity and started on the downward path.

While going to a lecture one Sunday by Prof. Felix Adler, of the Ethical Culture Society, I saw a man on the corner evidently suffering greatly from the cold. The saloons were closed and he probably would not have been a welcome guest at the meeting of the society, broad-minded though its aims are, even if he had been in a mood to listen to an abstruse lecture.

NO REAL SUFFERING AND DESTITUTION ON FARMS.

A short time ago I was on a visit to the South and I took particular care to note the conditions surrounding the living quarters everywhere. Perhaps the farmers of the South are not very prosperous; as a rule they do not appear to be as hard-working as those of northern latitudes; yet I saw no suffering such as even the most casual observer sees in any city. There are no

A LITTLE LAND AND A LIVING

half-blind children, no such degrading occupations as picking over slop-barrels!

It is clear that the woman with the child picking the refuse and being jostled by bumptious policemen, and the men looking for work at shoveling snow, would be much better off on a farm, and that it would be much better for them to be sent to the farm than kept in the large city to help along the unearned uncrement in the most unsanitary districts; that the families of workmen, who had only money enough to go into the slot-machine parlors and "nickel" theatres, but the men of which are of good ability to work, should, of their own volition, rather select the farm, and thus be enabled to come to the city once or twice a year and take in the better forms of amusements. And the fellow who, whatever his proclivities may have been, could not enjoy ethical expositions because he was not dressed well enough to enter the rooms of the society, would certainly be better off if he were on the farm, as would the entire community. The young and able-bodied men of our cities who idle away their time in gambling, drinking and riotous living, all at some one else's expense, in most cases, should be induced to take up the healthful, vigorous life of the farm—there to see accomplished the results of his labors, not only in a monetary sense, but in his improved physical well-being as well. Many of the workers in the crowded, filthy factories and sweatshops would get double the price of their labor if put in agriculture, at the same time improving their health, instead of marring it by working and sleeping among the remnants of their work, five, ten and sometimes fifteen or more human beings in one stuffy room.

THE EDUCATED AND INEXPERIENCED ADAPTED FOR AGRICULTURE.

Many park benches in the cities are filled in these years of much education with college and public school graduates, who, in the words of Mark Twain, " know everything but how to apply it." Much of their knowledge—which they do not find use for even in menial labor, often the only kind they can obtain—if, indeed, they are so fortunate as to obtain that—could find its application in modern agriculture. Their learning, instead of being a curse which has robbed them of the best years of their lives, would on a farm not only help them in their work, but be an endless gratification and pleasure to them, both in their work and in their recreation.

SCOPE FOR "KNOWLEDGE LEARNED OF SCHOOLS."

In spite of the fact that agriculture is unique among occupations, in that it can be engaged in without one's first attaining any particular experience, there is ample room for the knowledge learned of colleges—and, moreover, agricultural colleges cannot fail to benefit both the "present" and the "prospective" farmer. Many successful farmers of the present day, however, began farming with very little knowledge. They had, of course, to meet the occasional "joshing" of the country-bred, but they bore this in good part, and picked up knowledge and skill with the readiness and pluck of the tenderfoot of the Western plains. They have not found that an academic course of study or book-learning in advance was necessary for a start as a successful farmer, such as is required of the beginner in law, medicine, etc., but still they found that they were engaged in a profession where knowledge is power, just as in everything else.

TOIL AND PAYMENT OF FARM LIFE.

The average city dweller is afraid, perhaps, of the drudgery and long hours which undoubtedly prevail on the farm. But the drudgery on the farm is not like the drudgery of the city, which has a twin—suffering. It is a mistake to think that the farmer works longer hours at harder work than the man in the city. There are thousands of people in our large cities who would be infinitely better off, morally, physically and financially, making a living in the country by no harder work and without the misery involved in cheap living in the cities, where good, healthful foods are well-nigh beyond the meagre allowance of the workingman's scanty purse.

THE WORLD'S IDEAL FARMING COUNTRY.

The United States is the world's ideal farming country. Its varieties of climate and soils make it a natural producer of almost everything that can be furnished by the agricultural and stock-raising interests. There is not the slightest doubt that farm land will ultimately become the most valuable asset of our country. Benjamin Franklin said a century ago: "This country is fond of manufactories beyond their real value; for the true source of wealth is husbandry." Franklin was possibly wrong in his deprecation of factories in his day, for manufacturing has proved to be a prime factor in the present greatness of the United States. But he really spoke with the tongue of a prophet, for this is the time when the words of America's greatest philosopher on agriculture are pregnant with meaning and force. And this is likewise the time when this great manufacturing country should awaken to the terse and intensely direct sentence of Gibbon: "Agriculture is the foundation of manufactures."

AGRICULTURE'S GOLDEN AGE.

Those very manufactories have made possible the Golden Age of farming, which has now begun. No more remarkable inventions have been made during the last fifty years than those agricultural appliances which have taken the place of the sickle, the flail and the wind. What a difference is there in reaping, cradling, raking, binding and mowing, even in planting, from the days of old! The antiquated methods to succeed with which a man had to get up at four o'clock in the morning and work until long after "early candlelight," are responsible for much of the ignorance of city people in regard to the farm life of this day. What drudgery used to be entailed on the women of the farm, which is now done away with by all kinds of modern improvements in their work!

The modern farm-house, with its labor-saving appliances, its piano, its books and magazines, its compartively short hours of toil and its manifold interests, is an unknown quantity to the city woman, who thinks she is comfortable in a small flat, where she is under the dominion of the cook and the janitor. Many city people under present conditions, however, think themselves lucky to be under that rule.

The last twenty-five years have been the greatest transformation-period the world has ever known—and have brought as great changes in country as in city life. Even on farms remote from towns and cities, telephonic and telegraphic communication, and the later boon, rural free delivery—have come to keep their residents in close and constant communication with the outside world. These have annihilated distance for the farmer of to-day.

INDEPENDENCE IN AGRICULTURE.

Some of those who have guided the destiny of young men have realized that there is danger in a land that suffers from over-education and over-crowding of the professions. The careers of our young men plainly show that society is in need of both brains and brawn, and that while we have need of the professional class we must not neglect those callings that give us our bread. Ex-Governor Hoard, of Wisconsin, exclaimed in a public speech: " I cannot bear to go to my grave until I see imparted to my nation the spirit that shall make agriculture not only the support of men's bodies, but an inspiration to their intellects."

It has been ably shown in the recent reports of the Department of Agriculture that it is an inspiration to men's pocketbooks and bank accounts as well as to their bodies. Many who read these reports are surprised at the results of the years of unequalled prosperity. The immense annual increments, which are comparable only with astronomical figures in their demands upon the imagination, have a real significance when capitalized and expressed in terms of value based upon earning power. In one report the Department stated that in five years farms had gained in value by a third, or nearly $7,000,000,000, and that within those five years nearly 1,800 national banks had been organized, mostly in the Southern and North Central rural regions. These banks depend upon farmers for their business, and are not organized, as would have been the case a few years ago, with Eastern capital. "It has been an era of small banks in isolated communities," said the State Bank Commissioner of Kansas, at the time, "and so many have been started that to-day every hamlet in the State where any considerable business is done has a

A LITTLE LAND AND A LIVING

bank." This shows that the farmers are sharing in the national prosperity which they have done so much to bring about and are banking their surplus and looking for investments. How few of the city tenement dwellers can do that! I sincerely wish that these facts which the Agricultural Department collects could be placed in a striking way before every man, woman and child in the large cities of the United States, who might be induced to seek their fortune in a field than which nothing is surer of bringing full returns for honest work.

LAND ENOUGH FOR A LIVING.

Your book "Three Acres and Liberty," shows that with little work three acres will yield enough for any family to make a living on. If a man will devote all his time to it he might make a living for half a dozen families, using only so much as his family needs, and saving what would be the living expenses for five families.

ONE POINT OF HONEST DIFFERENCE.

There is one thing, however, in which I disagree with you. You believe that the city is more favorable to general education. It is my opinion that education comes more from reading, reflecting and the observation of nature than from mingling with crowds. The farmer has more opportunity to read and reflect, and less opportunity to talk, and consequently learns more than the average city dweller, whose mind is confused by conflicting ideas and schemes and whose time is spent in the hunt for amusement. The children could learn more where they could read, but where they would see less of the "high" life and degrading low morality of big city life, and be the better for it in after years.

WHO SHOULD GO BACK TO THE FARMS?

One of the most potent reasons that makes this a particularly good time to promote a vigorous increase of rural settlement, is the present high cost of living in the city and the destitution among the poor, ill-clad and ill-fed city workers.

But who shall go? There are thousands even of the most careful and industrious men and women in the cities now, who are making a bare living and nothing more; and who have always the fear of loss of steady work, such as is brought about under modern conditions by strikes, lockouts, and by improved machinery. The division and subdivision of labor and the constant change in cities in this age of specialization puts many out of employment—hence out of bread—when factories shut down. This surplus labor should go to the farm.

Those lads and young men who have left the farm for the city, and are not meeting with the success which they expected, should, in most cases, be encouraged to go back to the farm. Many young people continue in cities, suffering, in many cases, want and discomfort, simply because they are ashamed to return to the country after having, perhaps, boasted of their coming success in city life. They should be shown that it is no disgrace, in most cases, to have failed in the city. Or, if their lack of success has been due to the temptations of city life, they should be likewise induced to return to the farm, where those who have once yielded to temptation have better chances of growing up to be reputable and honorable citizens. The very city experience which young people who revert to farming life have gained will aid them in becoming progressive beyond former ambitions.

WHO WOULD BENEFIT BY SUCH A MOVEMENT?

In settling the question as to who would benefit most by the back to the land movement, we would also determine who should take the initiative in making the theory a practical success.

First of all, and in a large degree, the public would benefit, even though much of it remains in the cities; but, as Henry George said, "these bad conditions are due to the lethargy of the public." Consequently, it should be the business of philanthropists and sociologists to arouse interest among the public.

The next who would benefit (and possibly they should be placed first, because theirs would be chiefly a monetary benefit), are the railroads. For instance, if one hundred people working during the day in New York would live upon farms, the railroads would carry them to and from the city, and would haul their products also. In this way the railroads would make money out of the enterprise, once a colony was started.

SEMI-AGRICULTURAL COLONIES.

Heretofore railroad communication followed town building, and later town building followed the opening of the railways nearby. The railroads should look into this proposition and should co-operate, to the extent of supplying land along their lines, advancing the necessary funds to lay out and build, or going so far as to build houses themselves, renting the homes at a nominal figure and realizing on the passenger and freight traffic, which would increase year by year as the colony grows. In the end such an investment would bring handsome returns and would go far toward making the railroads a real big factor in civilization—not only in a commercial sense, as heretofore regarded; not only as avenues of communication between distant points; not only as builders and openers of empires, but also as builders of a new and healthful race, clutched from the fangs of the tens and hundreds of thousands of misery-dealing "reptiles" of the social and commercial life of our big cities.

The colonies, which I would call semi-agricultural, ought to be started near each city. The railroads should run convenient commutation trains, and soon the experiment would prove a success. Possibly some people would come back, but not without having had at least a good vacation for a year or two, and when they come back they will remember "the good meat they ate in Egypt." Movements somewhat similar have been tried by Socialists and economists, more than once, but they were confined to people imbued with certain teachings of their leaders. The new "back to the farm" converts would owe no allegiance to any particular party or sect and would follow the dictates of no person or persons, but, instead, would heed only the call in their hearts for a happy, healthful home for their families; their own patches of land and accounts in the savings bank.

I had some talks and correspondence about this proposition with railroad men. Some I

A LITTLE LAND AND A LIVING

"made" read your book, and here are a few extracts from the great number of letters received:

F. H. La Baume, Agricultural and Industrial Commissioner of the Norfolk and Western Railway, writes:

" . . . In no other book by a Sociologist have I found so much sound sense and so many facts marshaled for the promotion of the 'back to the land' movement and for intensive farming as in that of Mr. Bolton Hall— 'Three Acres and Liberty'—which he has sent me upon your suggestion. It certainly ought to convince even the most doubting Thomas that for a great many people who are now fretting along in large cities, hardly able to make both ends meet, the land of promise is not far away.

"I have seen Mr. Hall's theory—the 'Three Acre' proposition—fully vindicated, and I am proud to say that I have been instrumental in bringing about its practical application. Two years ago myself and friends divided a tract of land at Waverly, Va., by no means the 'most fertile' in the United States, nor with the 'most' favorable location, into 10, 15, and 25-acre farms, and upon each farm was built a three-room cottage. We originally sold one of these farms and cottages for $400, to be paid for in instalments, but this price had to be increased later to $500 on account of the increased cost of labor and building material.

"On these farms we have settled up to date about forty families, the men being of all vocations, but hardly one a farmer. Most of the ten acres are timbered, very little of it under cultivation, yet every one of these persons has made good not only in the payments to the company, but in saving something for themselves and increasing the value of their land by development work. In other words, by working only a very small part of the land which they have purchased for so little money (and only with cash payments averaging about $100) they have succeeded in establishing themselves in an occupation in which they have less worry and even less work than in many of the occupations that are offered in large cities.

"The hunger for land and a self-supporting home is inherent in every right-minded man. The natural habitat of mankind is in the country, and the city life at best is an artificial one. A reaction is setting in, and city people are to-day taking an interest in the country and its possibilities as a home that has not been manifested in years.

"The farmer certainly has the best of it. He works hard when necessary, but this individual effort brings direct returns in proportion, and though he may not deck himself in the fine raiment of his salaried brother in the city, his clothes are the clothes worn by the 'boss,' and he is respected accordingly. His hours of work are not longer as a rule than the city man's, and he has his Sundays and in addition many days when he is not compelled to be out attending to his crops. The winter season also affords him ample leisure in which to read, visit his neighbors, and keep in general touch with the world and its doings.

"He is not compelled to rely entirely on the butcher, the baker, and the groceryman for the necessities of life, but, to a very large extent, produces them himself. His garden furnishes him with an abundance of vegetables and small fruits, he has access to his eggs and tender young broilers without considering them a luxury. A few pigs keep him supplied with tender young pork, which, if necessary, he can divide with his neighbors, and in thus taking turn about and dividing with each other, they are able to have fresh meat throughout the year, to say nothing of the delicious hams and sides of bacon that are put in the smokehouse to contribute to his comfort throughout the winter. His orchard supplies him with the finest apples of many varieties and in the late fall when he has tucked them away in barrels in his cellar, along with a barrel or two of sweet cider, a few rows of golden pumpkins, Hubbard squash, and large piles of potatoes and turnips; when the contributions of the boys and girls—the popcorn from the fields and the nuts from the woods, which every farmer's child takes pride in supplying—have been deposited along with the rest; when the good wife and mother looks upon the long shelves swinging from the cellar rafters and sees the rows and rows of fruit jars containing the delicious preserves, jellies, canned fruits, and vegetables which bear such eloquent testimony to the loving thoughtfulness and preserving foresight of herself and her girls; when the farmer and his family, I say, sit together in the late fall and contemplate these many blessings; when the buzz of industry and contentment from the poultry yard, the stable, and the pen is adrift in the air and permeates everything; then with the thoughts of the Thanksgiving and Christmas approaching, the long winter evenings with their complement of cider, nuts, popcorn, and apples, the sleigh-rides and jolly gatherings at each others' homes—then the farmer and his family can indeed count themselves among the princes of this earth, secure in the conviction that the real blessings of life have been dealt out to them in generous measure.

"Now, let us look at the city man's side of the question. I mean the laboring man and the man who works for a wage. In times of prosperity, it is very true that he can command the conveniences and many of the luxuries of life; but prosperous times make high prices and therefore he is compelled to pay liberally for all he receives. If he has been fortunate and provident, he probably owns his own home, which affords him bodily shelter during times of depression and adversity, but is in no sense a revenue-producing investment. It is not necessary to go into all the details of the adversities, the mental and physical suffering endured by the wage earner in times of strikes, lockouts, and periods of business depression.

".The man who has a little farm can fall back upon it when other avenues of support fail him. It is

"1. A refuge in times when adversity, business stagnation, or strikes and lockouts deprive him of work or his usual source of support.

"2. A healthy outdoor life in God's own sunshine, where his sons and daughters can be reared in an atmosphere that will insure self-reliant, strong and morally and physically perfect men and women, a credit to their parents and to the country that gave them birth.

"3. The knowledge that he is an American freeholder, with all of his privileges and responsibilities, and therefore, is as vitally interested in the affairs of this great country and as fully entitled to her protection and consideration as any other citizen in the land.

"4. Last and best of all, the knowledge that he has a comfortable and productive home awaiting him in his old age, where he can be assured of a competence and a refuge after he has reached the pinnacle of life and is traveling down the farther slope over a path which has been strewn with many blessings due to the foresight of his younger days. Near every city which offers a ready market for the product of the farmer or the " amateur " farmer there is land which could be satisfactorily and profitably used for three acre farm plots, and it will be a blessing to this great country if Mr. Hall's theory can be tested soon, for each test will more than show its practical side. We have on this tract people from all walks of life—clerks, mechanics, professional men, those previously engaged in all trades and professions. All of them are less than two years on our tracts, but you will not find anywhere else such a happy lot of people are found there now.

" I enclose you a few photographs of those farm houses which I take the liberty to call ' cottages.' Compare the exterior with tenement houses surrounded by other tenement houses and these little ' boxes ' surrounded by gardens and pure air. Compare their interiors with the interiors of the tenements, without light though possibly with more elaborate furniture, and you will find that there are thousands who could be made happy by going to just such places near large cities, even if they have to go to the city to work during the day."

OTHER LETTERS FROM RAILROAD MEN.

Extract from a letter from an enthusiastic officer of a prominent railroad:

" . . . I have resigned my position here to take effect on or before May 1. I am sick and tired of the city, and want to get in the country to spend my declining years. I will see what can be done about getting the necessary land for the colony you speak of. I would gladly give it myself, if I had it."

This is from a man in middle age, hale and hearty in body and also in soul and mind, or he would not see things as they are.

Mr. Fred P. Fox, industrial agent of the Delaware, Lackawanna and Western Railroad Co., writes me as follows:

"One of the most serious problems of the present day is the reclamation and rehabilitation of unused farm lands, especially east of the Mississippi River. The subject divides itself into two questions:

"How to get the youth of the country, bred on the farm, to stay there?

"How to induce the city bred youth to take up farming as a profession?

"These are questions that are occupying the attention of many great minds.

"For the country-bred youth, better schools, teaching the chemistry of agriculture, the technique of farming, the analysis of soils, better environments, etc.

"To the city-bred youth, literature telling in pleasing form the profits to be derived from farming in money as well as the larger dividends of health and happiness.

"In regard to the appeal to the dwellers in cities the

A LITTLE LAND AND A LIVING

best message is contained in Mr. Bolton Hall's 'Three Acres and Liberty,' which you so kindly sent me with your compliments. It is as fascinating as a story, as true as history.

"Down deep in the hearts of all there is a desire to spend the last golden days of earth on a little plot of ground in the country. It is the picture drawn by Hope and colored by Faith, that makes life worth living in the congested marts of trade.

"Some day, somewhere, I'm going to have a little home in the country'—nine out of ten people say this, believing it will come true.

"Railroad managers are deeply interested in this problem, for next to a paying industry along its rails, should be a paying farm, for without the farm there could be no paying factory. The Industrial Departments of the railroads have on file and are interested in giving out literature, maps, and particulars in regard to vacant farms along their lines.

"The idea of making these desolate gardens of earth bloom again like the rose; to see the abandoned farm houses filled again with bright-eyed, laughing children; to see deserted fields covered with flocks and herds; to see again the golden harvest days, is a consummation devoutly to be wished, and there is no more blessed mission than bringing it about."

Still another enthusiastic note is struck by Mr. Ira H. Shoemaker, industrial agent of the Delaware and Hudson Railroad:

" . . . I am in hearty accord with the ideas which

are so attractively presented by your friend, Mr. Bolton Hall, in 'Three Acres and Liberty.' I am much interested both personally and as a railroad man in the possibilities of support to be derived from intelligent, intensive farming applied to small land holdings. People who have wearied of 'flat-cramped, closet-roomed apartments, or pigeon-holes for humanity' will find the helpful suggestions of this book invaluable in aiding them to find real homes which will not only contribute to their health and pleasure, but also to their material support.

"In studying the industrial conditions of our own railroad and other roads, I am fully convinced that the value of tonnage is greater from one acre of land intensively and intelligently farmed than from large tracts cultivated in a careless and inefficient manner. Therefore aside from any altruistic motive, I hope that this book will create a widespread interest and that many will be induced to put Mr. Hall's ideas to a practical test."

These letters show plainly that these people see the benefit, but they seem to be afraid of the slowness of the general public, or they would start active work in the three-acre direction.

A PLEASING TENDENCY.

It is a pleasure to note that there is now a tendency among those who have made money in the city, or those who are in good positions, to take their families to spend the remainder of their lives amid the rejuvenating influences of the country. These men come to the city daily to work, yet when you talk to them about their efforts in the agricultural line they fairly swell up with pride and without a shade of modesty they will tell you of the wonderful specimens their little gardens produce. These gardens furnish them as much gratification as their life-business. Business men, as they feel able to withdraw from the severe activities of commerce, are looking forward to a happy old age in the rural districts. Many fine farms are already being developed by these men of wealth. The movement is developing among this class to a surprising extent and also among the brain workers of the commercial centres, and you have only to visit the railroad stations or the ferry houses, morning and evening, to get an adequate idea of the number of business men who come daily from their country homes to their work in the city. As these people buy farms, land will become more expensive, and for this reason alone, now is the time for the man of small means to go into farming.

THE LETTER THAT PROMPTED THIS BOOK

WHAT I HAVE DONE.

Aside from talking in my own circle about the proposition I addressed an open letter to the Secretary of Agriculture early in 1906, and more than 500 newspapers in the country, including many of the leading ones, noticed it. It elicited editorials from such men as Henry Watterson, Lafayette Young and others. Giants of the press have realized the national importance of the subject. The suggestions I made to the Secretary to promote this movement were novel yet feasible. My appeal to have the Government, through the Department of Agriculture, facilitate matters by providing special bureaus and publicity to acquaint city dwellers with the golden opportunities awaiting the worker in agricultural fields has, as yet, been unheeded. Mr. Wilson received the letter, yet nothing has come of it, possibly because the Honorable Secretary realizes the inertia of the public. However, he has given the idea a general endorsement, for in receiving a delegation which came to interest him in a plan to open lands to the unemployed, he said: "The duty and burden of establishing the enterprises rest unmistakably upon the State and municipal governments. The great manufacturing centres, like New York, Chicago and St. Louis, have had the advantage of large populations in time of industrial prosperity. They have drained the country district of a large proportion of the wealth-producers, and it is only fair that, having had the advantages, they should shoulder some of the disadvantages. They must not expect to eat their cake and have it, too.

"The idea is a good one, and the distance between a closed-down factory and a newly-plowed farm should be made as short as possible. Let the great municipalities, and the smaller ones also, for that matter, provide land within a five-cent fare from town; let the unemployed have access to it, and we will send them experts, without charge, from the Department of Agriculture to show them how to do it, and we will furnish the seeds also. It would solve the problem of unemployment as quickly as anything else would, and the worst that could happen, in case of a return of prosperity in the factories, would be the possession of valuable farms by the municipalities, which they might conduct profitably on ordinary business principles."

WRITE A NEW BOOK!

"To the making of books there is no end," but it is books, and particularly such practical ones as yours, that educate the public to the needs of their conditions. I should suggest that you write another book on this subject. It must be shorter in order that even he who is busy would start to peruse it; that even "he who runs may read." That book ought to sell for little money. It ought to notice your "Three Acres and Liberty," so that those who read the smaller book and find that they are interested, and, like Oliver Twist, want more, should get the first book as a second dose—the heavier one.

The new book should answer the old question: "Who are they that shall go?" And if the public will but take notice and heed, they will give the answer that Moses did: "We will go with our young and with our old; with our sons and with our daughters; with our flocks and with our herds we will go."

YOU, MR. HALL,

are a man of strong powers of observation and have the ability to express your thoughts in an inspiring manner; you know the sociological conditions in this country as well as any man. Can you not make glow to a flame the spark that is existent in everybody? You hear from a great number of people that when they get a competency they want to have a spot of land that they can call their own and where their wives and children can ENJOY life. This shows that it is human nature everywhere for every one to long to be nearer Mother Earth and to claim a patch of the land as his own.

In your book, "Three Acres and Liberty," you have shown one side—that it pays in money to go back to the country and take up agriculture. Now write a book which will show that it pays from other standpoints as well. You do not have to go far out of your way; the means are near at hand and the evidences for the movement are to be found everywhere you turn in the cities. Let the people go to the land for their own sake, for the good of their children, and for the good of others that remain in the city; for the good of the whole country, and, last but not least, for the good of the human race. If a few go away from a tenement house, it will give a chance for only five to sleep in a room where ten are now crowded together. Let a few of the unemployed take up duties on a farm, there to earn wages and live happily, for then only fifty men would apply for a position where now a hundred apply.

THE CAUSE AND THE REMEDY.

The tendency of population to flock to the already congested cities is a menace to the prosperity of America. While much of the brain and nerve power which is so great a force in the cities of this country was originally nurtured on the farm, the time has come when the farm needs to retain much of that ability which it has heretofore given so profusely to the city. The twentieth century will recognize the farmer as the king among the workers who benefit mankind. The dignity and independence of that which our first President (himself a farmer) called "the noblest occupation of man," will be illustrated in this country as never before. The young man and woman will find it profitable as well as pleasurable to stick to the farms, or to leave the crowded cities for the country, where one with small capital may more easily secure credit and a competency, and can have better opportunities for the right kind of "life" than that which is concomitant with limited means in the city. The superiority of country life for raising families will be recognized by all, and the happiness of being "near to Nature's heart" will be truer and better than the pleasure which the city affords.

FOREIGNERS LIVE MORE CHEAPLY.

In urging this "back to the land" proposition, I have purposely omitted any discussion of the foreign element and the influx of immigrants. It is noticeable that, in spite of the congested conditions of the foreign districts of our cities, and the almost pauperous conditions in which they seem to live, there is not that extreme distress and destitution which prevails in the other poorer districts of the cities. This is probably due to the fact that foreigners, generally, live more simply than our poor. As they live their own characteristic life, and are thrifty, they make their scant earnings from vending, sweat-shop work, or that of their own little shops, or in their hard out-door labor, go twice as far as those of the other class of poor, with their unwarranted perverted tastes for "high life" and extravagant wants.

However, in the congested Jewish and Italian quarters, with their millions of new comers each year, there lies a great problem, and it is gratifying to note that it is being solved, though slowly to be sure, by organizations of their own better-situated countrymen.

HOW TO MEET THE INFLUX OF POPULATION.

How are we to meet this influx of population in a way that will be of the greatest benefit to the increasing hordes and also to the United States? We do not intend to deport such of these immigrants as are mentally, physically and morally sound. That is not the spirit of America. Many of them come to these shores with the intention of taking up farm life, but get caught in the maelstrom of city life. Several societies are making systematic efforts to send as many as possible to farming communities where labor is needed, or to buy lands and lay out farms where families released from the sweatshops and tenements of the cities can live in comparative comfort while becoming independent and earning homes. Such experimental colonies have been started in small ways by the Baron de Hirsch fund, the Industrial Removal Society, the Jewish benevolent societies and others. For instance, at Arpin, in the northern woods of Wisconsin, a farming settlement has been in operation for over three years and the plan seems to be working successfully. The ground was laid out and each family was provided with a team of horses, a cow and necessary farming implements. The men were paid a weekly wage by the association and were on probation for one year. At the end of the year, if mutually satisfied, a contract was signed, giving the family ten years in which to pay for the land and equipment, and the profit realized by the association from the man's work over the wages advanced was applied to the first payment. In spite of poor crop conditions, the majority of the original eighteen families remained throughout the first year and the quitters were quickly replaced. It is estimated that it took an average of $1200 to establish a family.

CONCERTED ACTION NECESSARY.

All these efforts, you will readily see, are as only a drop in the bucket. They should have the aid of an arm more far-reaching, with unlimited resources and facilities, so that all societies and individuals might unite their efforts in one big organized movement for the welfare of the poor and unemployed of our large cities whatever their nationality and religion. By concerted action alone can permanent and influential results be accomplished; otherwise while the process would go along slowly, conditions in the cities would grow worse and worse, without hope of our good efforts catching up with them. The United States Government should follow the example of Canada, and even go further in influencing the poor and unemployed to go to the land, by holding out special inducements.

Briefly, I would suggest that such proofs as can be obtained about the condition of agriculture in this country should be widely disseminated; that the Department of Agriculture establish, in each of the large cities of the country, offices which would give full information to inquirers who wish to know of the best places in any part of the rural districts of the United States in which to take up their residences with a view to engaging in agricultural pursuits. These offices should, I believe, co-operate with associations and with individuals to draw away from the city such elements of the population as are not likely to succeed there. This could be done through co-operaton with charitable associations, immigration officials, lecturers, the press, etc.

If results from this educational campaign do not materialize within a reasonable time, then there should be more widespread and drastic measures, even going so far as supplying the land, and even the home, from the government's own tracts or from the various States' abandoned homesteads. Yea, I would even go so far as to compel to go to the farm those who are manifestly unable to eke out a living in the city and thus are on the way (if they have not been so already) to injuring themselves and becoming a charge to the community. Ultimately the community, for its own protection, sends its vagrants and criminals to the workhouse and the penitentiary, and it is largely from the ranks of these unfit citizens that criminals are made. Why not prevention instead of a "cure" that will never cure!

MANY MEN OF MANY MINDS.

To-day there are the most conflicting ideas on foot to aid the poor and needy of our cities—some practical, most of them merely theoretical—but too many for the good they may do.

Ex-President Cleveland and Prof. Felix Adler seem to realize, as the Bible says, that "the poor shall never cease out of the land." It was Prof. Adler who a few days ago declared boldly that there was no hope for the poor; while Mr. Cleveland, in an interview in a daily paper, holds out only a faint hope for the needy ones, and wishes to encourage charity through the medium of a sort of clearing house or charity trust. On the other hand, there are a great number of people who would like to relieve the situation—I happen to be one of them—by giving the poor what is due them, or, in other words, by not stealing from them what is their inherent right and so keeping them poor. Yet in the light of to-day such philosophy gives little hope —a theory that will not be practically applied before the people will realize that nothing is holy because it is hoary, as well expressed by the words of Schiller's "Wallenstein":

> "Woe unto him who dares to shake the ancient,
>   Dignified run of things, the precious heirloom of his
>   sires!
>   Time does exert a mighty soothing power
>   On what hoary is with age that is for him divine.
>   Be in possession and thou dwellest in law and right,
>   And sacredly the crowd will recognize what thou
>   possess."

THE REAL HOPE.

The idea which you so strongly and ably expound, Mr. Hall, is the only real hope—the one ray of sunshine that is destined to lighten the burden and cheer the lives of the needy of our large cities. The movement to get the poor to the farm will help a great number, if not all, and the results will be permanent. However much charity would help, it keeps only for a short time, and it has its drawbacks. As the honorable ex-President says: "Charity is responsible for the apparent contented dependence of the poor." This dependence, he says further, "is the thief of self-respect. The poor are not permitted to help themselves. [It is the privilege of the few to help themselves with that which belongs to the many.] Once they come within the ken of the charity workers they are robbed of their initiative and are made dependent."

Mr. Cleveland's remarks bring back to me very strongly my contention that of all the plans to aid the poor yours is the ideal one. It would do away with all the drawbacks of the charitable aid and promote independence instead of dependence—not unlike what Macaulay pictured:

> "Then none was for a party;
> Then all were for the State;
> Then the great men helped the poor,
> And the poor man loved the great;

> Then lands were fairly portioned . . .
> The Romans were like brothers,
>   In the brave days of old."

The time has arrived when the increase of education will send people back to the country, and will populate the rural districts with those who will apply their education to a profession which is becoming more and more profitable because of the dependence of the cities upon agricultural districts.

By writing "Three Acres and Liberty" you have done a signal service to your country and to your fellow men. Our editors appreciated it greatly, as was shown in the notices the book received, which must bring home to you your conviction that all these efforts are for a good cause. Now complete the work which you have so ably begun, and we will see a mighty concerted effort to make the movement to the land a practical success. You have pointed out the way, now go on, write another book, that will discover the leader.

Yours sincerely,
WM. BORSODI.

New York, March 18, 1908.

A
LITTLE LAND
AND
A LIVING

CHAPTER I

LIFE, NOT MERELY MAKING A LIVING

Opportunities Past and Present—Chances Lost—New Chance—Intelligence—Slums a Symptom—The Origin of Wealth—Capital—An Acre Enough—Possibilities—Thoroughness—Available Land—New Methods—Intensive Farming—Trucking—Dairying—How They Do in Japan—Denmark—Jersey—We Can Do Better.

WHEN a goose goes under an arch she ducks her head; that is not because there is not space for her, but because she thinks there is not, and that is because she is a goose. Perhaps she does not see very clearly what is above her.

But we can hold up our heads if we will; it is only necessary to look and think. Many of us believe we had no chance in life. Others were in at the oil excitement or the natural gas boom or got in with the trolley development or a mine discovery, or found some good opening; but we have never seen any opportunity.

Well, we have our chance now. A new boom is on, the farm land boom; a new development is beginning, intensive agriculture; a new discovery, the riches of the soil; a new opening, the intelligent use of "the little lands." It does not demand any more brains than any of the other opportunities and it is open to a far larger number. "The profit of the earth is for all."

There is more money to be made out of the soil, if you go at it intelligently, than there is in any endeavor that is open to everyone. The city man who has brains enough to conduct a shop, or who knows how to make a profit out of his employees, or who is a good enough fianancier to meet the monthly bills, knows enough to make money out of the soil. The same attention to details, the same care of an orchard or of a vegetable farm or of a fruit farm or a flower garden, will bring far greater profits. Anyone who has a little store or who makes things in a small way, is oppressed with the ever present danger of being crushed by a trust or forced to the wall by richer and more powerful competitors. What chance has a woman in the city now, other than a mere living? And what chance has the average clerk? Both grow old trying to keep abreast of their expenses. Many of these people have a natural taste for the land. They love to prune, they love to plant, they love to help nature perform her marvels. They potter away in their little garden patches at their homes in the suburbs, and find more real enjoyment in their gardens than in anything else. These are the ones who might make grand successes of their lives if they worked the soil. And it is a life worth living, with real social advantages.

Civilization does not breed over-crowded cities, nor do we need to stay in them. The slums and the billionaires are not diseases, but the symptoms of a disease, the divorce of the people from the land.

All things that go to make well-being, things that we call wealth, come from the land by work; so that all political economists agree that wealth is anything that people want which is gotten out of the earth by labor. Cotton, wool, linen, leather, felt, all that we wear; wheat, fish, beef, vegetables, everything that we use, comes directly or indirectly out of parts of the earth.

Even the money that we buy or sell these things for, the tools that we use to make them, the machines with which we manufacture them, are themselves drawn from the land, so that economists say again that capital is only that part of wealth, of the product of land and labor, which is used to make or to get more wealth.

To get the simplest kinds of these things in the simplest way, and to let the children learn to get them in the best ways, a few acres is enough, with modern methods and active minds.

An acre is not a big plot; the base ball diamond has about a fifth of an acre. About two blocks of the space on Fifth Avenue, between the houses, is an acre.

On such a plot any dunce can raise the average crop of onions, say three hundred bushels, and make at least big wages. If we use brains, and transplant from hot beds, we can raise that yield to five hundred and fifty bushels of choice early onions. That is much better and more profitable than "to farm" and raise four hundred and fifty bushels of rye on 80 acres, and it gives much more satisfaction. The farmer's work, like the woman's work, is never done, and what gets done is never done thoroughly. Toil as he will, there is still crying need for more work at the house or somewhere on the fields until he gets so worn and discouraged that he has not time or the energy even to make a shelter for his machine or to take his tools in out of the rain.

But just such thoroughness, just that kind of care is necessary to success in any business; for the husbandman it is practicable only on the little plot, or what is practically a lot of little plots.

The successful grocer and the successful cotton spinner start their business near their customers, or near their labor; we must start our garden or our little onion farm in the same way. The price of land will quickly teach us that we can get but little of such land, and that therefore we must use it to the very best advantage.

For that is what the best department stores and the best bankers and wholesalers do; they get the best possible situated bit of land regardless of the price, and put it to the best possible use. But the farmer is often working on a place that is literally worth less than nothing, for the whole outfit will generally sell for less than the fences, drains, buildings, and the "good bearing condition to which the land has been brought" could be reproduced for.

So it will pay us better to hire or buy an acre or two, worth a thousand dollars, near the market, than a farm for fifty dollars an acre far from the market. It saves capital, increases returns, lessens risks, and facilitates education to be near the city on a little bit of earth, for it can be rented without a fortune, worked without buildings, and its product can be sold without delay. In primitive times when we had virgin soil which cost little or nothing, a more evenly distributed population and little foreign competition, the farmer could get "independent rich", raising varied crops on a large area. To-day, with the high price of farm lands and the control of the markets by the railroads, which will often haul freight to New York from Chicago as cheaply as from Albany, it is well nigh impossible to get more than a living that way

Intensive farming, intensive trucking, intensive dairying, and specializing are the up-to-date methods that promise sure and good returns.

Three cows to the acre, not three acres to the

cow, $700 produce to the acre, not seven acres to the hundred dollars produce; four truck crops to the acre, not four acres to the truck crop—these are the methods that pay.

This is no "pipe dream" nor experiment. Japan shows us communities living off little patches of land of two and a half acres to the family. Denmark is a country of prosperous little farms. The Island of Jersey in the English Channel is about eight miles average length, and less than six miles wide, about the size of Staten Island, in New York Harbor; high rocky cliffs bound its coast on the north and west. Its agriculture sustains its 60,000 inhabitants, more than three to the acre of arable land, which is in the hands of about 2500 owners, who get a yearly rental of $50 to $100 per acre. The soil is not very fertile, but its productiveness is enhanced by mildness of climate. The holdings vary from three to thirty acres, and herds of more than a dozen cows are very rare; land and grass are so valuable that cattle are never permitted to roam at large, but are all tethered and are moved several times a day. They are always led by women instead of being driven. The cows remain out of doors the greater part of the year. Very little grain is fed, but in addition to grass and hay, the cattle are liberally supplied with roots, chiefly parsnips. With our higher intelligence, lower rents, and better machines we can do much better in this country.

"The Man with the Hoe" is a back number; it has been proved by measurement and by clock over and over again that a man with an ordinary hand cultivator costing less than ten dollars, and lasting for years, can do as much work as ten men with hoes.

You remember the story of Hercules and the wrestler Antaeus: Exhausted by fierce competition, Antaeus gained new strength every time his feet touched the ground; he was overcome only when he was lifted bodily from the soil. Get your feet upon the soil literally and figuratively, and draw from it not only wealth but health and the joy of the earth—not only a living, but Life.

Earth is your Mother; honor her that your days may be long in the land that the Lord thy God giveth, for all the children of men.

CHAPTER II

BUYING A GARDEN

Ownership—Sources of Information—Agents—Where Not to Go—Fertility Secondary—Soil Never Barren—What to Buy—Run-down Farms—Cheapness Relative—Where to Buy—How to Buy—Don't Fear Debt—Borrowing Money—Insurance—The Monopoly—Unproductive Lots—Use and "Improvements"—Make Unused Land Pay—To Buy or to Build.

IT is necessary to get hold of a bit of this earth; even a lease is a limited ownership, but it is far better to buy.

Do not be tempted off to the deserts or the tropics; thousands of years have been used in adapting your mind and body to contact with your fellows and to temperate climes. New conditions elsewhere are uncertain. Western Kansas and Nebraska were settled up after 1883 and a few wet years gave fine crops; then a few dry years and the grasshoppers brought despair, and homes and thriving towns were absolutely deserted. With modern methods they are now coming into use again.

The State Agricultural Department will tell you a lot about climate and soils and general prospects and prices, and sometimes even of "abandoned farms," which you can sometimes get for less than the mortgages on them, and the United States Department at Washington will furnish you a map giving the soil survey and will tell you all about temperature and rainfall of most districts. The Industrial Agents of the Railroads will give you more points, but the big Real Estate Agencies are the ones you must go to for personal and particular directions. Find out all you can by writing, and especially by talking to anyone whose knowledge or even opinion is likely to be worth something. When you want anything, the way to get it is to tell everyone you meet what you want.

Don't try to skimp by seeking sellers yourself, so as to save the agent's commission; even a poor agent knows far more about what is to be sold and for how much, and what is cheap, than you are likely to learn by days of travel and weeks of inquiry, and he can generally buy cheaper and get better terms than you can, because he knows what arguments to use and will use many that, for the sake of your soul, you would not use.

Get a good map as soon as you decide what locality you like; then get from one of the banks the name of a real estate agent whom they can recommend—at least he won't be a known thief, as some are; examine one farm in that section and talk with those who have cultivated it, even in their crude way; by it you can judge of other land near by.

Don't be deluded by five-acre plots on easy terms, even if the soil is fertile, *where there is no market;* the farther you go from a good market the less the acre is worth. For a small piece of land, situation and cheap transportation facilities are of far more importance than is mere fertility; the German idea is, that a good soil is any place where one can put fertilizers.

You should look for a good soil, but situation is the thing. Location, nearness to a good place to get manure and to sell crops, has more to do

A LITTLE LAND AND A LIVING

with profitable cultivation than the quality of the soil or anything else.

The following extract from the Year Book of the Department of Agriculture for 1902 (page 562) shows what can be done with barren soil: "The Connecticut Agricultural Experiment Station has grown carnations and other crops in sifted bituminous coal ashes with three per cent. of peat moss. There is practically no plant food there, so for 100 square feet of bench-space six pounds and three-quarters of nitrate bone black, and muriate of potash were thoroughly incorporated with the ashes before setting the plants and proved to be better in some ways than rich compost."

So look out for a good place where you can get stable manure cheap. Don't be discouraged by the run-down look of a place or the ratty looking shacks—those are the places that sell for much less than they are worth. A lot of pains and a little paint will do wonders to renew a tumble-down, weed-grown farm, and you will get big pay for them. You are out for a bargain, and if a place looks well it sells well.

Get land that is in the line of improvements, so that the value will grow through the efforts of others while you sleep; it is cheaper to pay several hundred dollars for an acre that is sure to double in value because someone will need it, than to get a tract out at "jump-off" for a song.

If possible buy or get a long lease near a growing city and with good trolley or railroad connections. If you don't, the growth of the city will benefit only the land owner, and will in time crowd you out. If you buy, the increased value due to the growth of the city will be added to the profits of your crops and later will pay you for moving out of the way.

Get your land in a district that people are beginning to go to; there is where you can earn a living off the plot while it is advancing in value, but you must not buy more than you can carry through the hardest times, or through a time of sickness.

Have as much as possible left on mortgage for as long a time as possible, say five or even ten years, with the privilege to you to pay off earlier on any interest day. Of course if you can pay cash and have enough capital left, it is well, but don't be afraid to go moderately in debt; to borrow money to spend is one thing; to borrow it to invest is quite another. All banks and all merchants begin by getting in debt and adding the use of the capital so acquired to their own —even cash houses always owe their clerks.

The experience of the Building Loan Associations in the East, and of the pioneers in the West, shows that you may borrow money even at twenty per cent. and make money. It is not debt but recklessness and improvidence that ruins men. Invest carefully in something that you yourself know as much about as anybody else does—not in wild-cat stocks. Says Frederick F. Ayer, "Wall Street is a trap."

Insure your life and your health too, if you can, so as to make provision for those dependent on your efforts and get on your feet as soon as possible. You may get rich having others work for you. You never will by working for others.

But don't get excited over a "snap" and pay too much. There are as good chances to-day to get rich off land owning as there ever were; and there will be just as good chances to-morrow.

So many great fortunes are due to the rise in land values that we are apt to try to get rich that way without due knowledge of principles or of circumstances; and so to lose. Only luck can make up for lack of foresight, discretion, and experience. Lambs are shorn in the real estate market as well as in the stock exchange.

Remember in your buying that the ownership of land is a method, not only of making money by industry, but of making money by law. It is the fundamental monopoly, and intensive use of the land will enable you to hold it and live by it while you wait for the natural increase in value.

If you are one of the many thousands who have bought "lots" that have not risen in value as was expected, think whether they might not be made productive by raising vegetables or animals, instead of raising prices.

It is useless to wait for large advances in lots which are partly settled with small homes; few can pay over five hundred dollars for a home lot. Unless the home lots are wanted for business, apartments, or other expensive improvements, they have a very definite limit in value, because the small home owner does not greatly enhance land values and he will not move.

Determine what you want to do, lest you fall between two stools; whether to speculate in land, making a living off it while it increases in value, or to get produce and take the chance of increase in value.

If you merely want a place to work, nearness to an asylum, hospital, or institution will often help greatly, but such "bad improvements" hurt the speculative value as much as a cemetery does —unless it is so situated that a growing concern must have it. But if you wish to speculate— that is another story.

You can often buy a tract that the owner won't divide, and sell off part of it at an advance. If you have to carry land that you cannot use to its full capacity, see that there is some kind of shanty on it so that someone else will be working to pay you rent.

Look out for the probabilities of assessments for street sewers, sidewalks, and especially for grading. Get an old building if you can, even if you have to have it moved a long distance on to your land. It is cheaper to buy and alter by your own work than to build; a tumble-down barn with good lumber in it makes a surprisingly good house at a low figure, especially if you can do the most of the work yourself. I have tried it repeatedly and know, though I did not do the work myself.

If you have any special knowledge of soils, of minerals, of coal, or clays, or water power, you may make special profit of it. Thousands have got fortunes (as Carnegie got his first start by buying natural gas land) through seeing capacities or uses of earth that others overlooked.

CHAPTER III

VACANT LOT GARDENING

Charity vs. Self-Help—Opportunities—Assistance—Cooperation—Little Government Needed—Cost—Health and Success—A Variety of Returns—One-half Acre—Large Profits—Land, not Capital, Necessary.

> No nation was ever overthrown by its farmers. Chaldea and Egypt, Greece and Rome, grew rotten and ripe for destruction not in the fields but in the narrow lanes and crowded city streets, and in the palaces of their nobility. So let us thank God and take courage as we see in our day the movement countryward, and the "abandoned farm" and lot no longer abandoned. Surely the history of creation is repeating itself, and again is the Lord God taking man and putting him in a garden to dress it and keep it.
>
> Dr. Francis E. Clark.

WHAT unskilled labor with little capital can do on quarter acre plots of poor soil, well situated, is best shown by the Vacant Lot Gardening Associations of various cities.

Placing the half sick, the disabled, worn-out people and the unemployed on vacant lots, where they can employ themselves raising their own food, is now no experiment, but an important support of many families which would otherwise be dependent upon charity.

Captain Gardner, U. S. A., the first Superintendent of the Detroit Vacant Lot Gardens, in a letter said: "It is the opportunity to help themselves that these people want, and it does seem so wrong that in cities people should almost die of starvation and yet thousands of acres held for speculation lie idle within the city limits. It's a sort of a 'dog in the manger' business. Poor people are often as sensitive about being objects of charity as you or I would be, and as a rule they prefer to work for what they get— rather Vacant Lot work—than to receive it for nothing."

It has met with marked success between the waves of speculation, in over twenty cities throughout the United States and also in England and France, as a means of opening employment to those who are incapable of earning a living elsewhere. Its practicability and efficiency have been recently demonstrated, particularly in Philadelphia, where for years from one to two hundred acres have been kept in cultivation. In

1907 over eight hundred families raised on about two hundred acres, crops worth $40,000.

There are plenty in New York City who want land, but Vacant Lot Gardening there has a serious drawback—lack of land within easy reach of congested centres, that can be had free; however, thirty acres in the Bronx was loaned by the trustees of the Astor Estate. Only fifteen was cultivable, the rest being woods; the land was so rough that sod and stone, making ridges three or four feet high, were removed by the gardeners before the soil was in condition to plant.

An announcement in the papers brought in a few days more than three hundred applications, coming from the five Boroughs and some from Jersey City and from Newark. Applicants were sent to the manager at the Farm to secure a definite plot varying in size from an ordinary city lot to one-half acre. Such persons or families as needed assistance were supplied with seeds and tools by the Association.

A hotbed was made to supply early plants. An expert gardener was employed in the beginning to instruct the gardeners in preparing the soil, selecting and planting the seeds, cultivation, etc.; each gardener was expected to cultivate his own plot, and was held responsible for its condition.

In our gardens each cultivator received a card like this:

1. Each person receiving a cultivating privilege is required to cultivate the land throughout the season.

2. Each gardener accepting a privilege, agrees not to trespass on another's garden, but to co-operate with all in preventing trespass.

3. Failure to comply with these regulations will cause forfeiture of the privilege.

4. The decision of the Superintendent shall in all cases be final.

These simple rules have been found sufficient. Although the cultivators were recruited from many nationalities—French, German, Swedish, Italian, English, Jews, native Americans, Negro—there was not a single case of disorder and hardly any thieving.

It is hoped so to train these families in agriculture, that they can transplant themselves to the country, and so be permanently benefited, as has been done at Philadelphia, where in 1904 sixteen families who had gardens in 1903, leased nine acres at $15 per acre per year and made one of their number manager. Their children from nine to twelve years old sold the products to consumers; organizing their own routes, and receiving daily twenty per cent. of their sales in payment. They often made four to five dollars a week each, never working over five hours daily, and at work that seemed to them more like play.

Such results encouraged the New York Committee. In New York the cost to contributors is about ten dollars per family. Against this the families have products of nearly $10 for every dollar expended.

The opportunity to cultivate was especially welcome to wage earners whose large families found no room for healthy activity in the narrow streets of Manhattan. A single man, too old for active work, had one of the best gardens. A consumptive, aided by his wife and two small children, cultivated an eighth of an acre, producing fifty dollars worth of products. This man lived in a tent all summer and built a shack

12 x 20 feet for the winter. Although not cured, his condition was much improved. Another man, a sailor, who was a nervous wreck as the result of an operation, performed a prodigious amount of labor on a piece of the roughest land on the farm. His condition was such that he could work but a few hours at a time, thus being unfitted for any steady employment. Nevertheless, his garden compared favorably with any of the others. A Jewish family, mother and eight children, worked assiduously on a half-acre plot, the largest granted to any one family.

Some, who lived in the vicinity, did the planting and weeding early in the morning and after work hours in the evening. Others residing far down in Manhattan, came to their Gardens Saturday afternoons and Sundays. A few hours during the week was sufficient to cultivate one-quarter acre throughout the season. The time required for one-quarter acre is about 72 hours, in a season covering six months or 24 weeks. This gives the time required per week as three hours.

The product of one-quarter acre varies with

the skill of the gardener. Our experience bears out the results that have been attained in Philadelphia—to wit—a product of fifty to one hundred dollars per garden. For the time put in, the garden earns, therefore, from 70 cents to $1.40 per hour—without any of the advantages of capital and without any employer.

In addition to the product of the garden, those families who lived in tents saved from fifteen to twenty dollars per month in rent. For six months the entire saving amounted to from $140 to $220 per family. The expense was the cost of the tent—about $20 complete—and such items as seeds and tools, which for a quarter-acre garden amount to five dollars for the season.

The indirect benefits to persons with large families were very great. In a few weeks after going to the Farm the pale, puny children became ruddy and robust, playing in the grass and living healthy, natural lives. They helped the mother in the gardens, and added their mite of strength to weeding the growing vegetables, to feed the family during the summer and to make a store for the winter.

A LITTLE LAND AND A LIVING

Ignorant people, women, cripples, boys, the aged and infirm have proven their ability to support themselves from lots 100 x 150 feet of poor land without hotbeds, greenhouse or any permanent improvements.

Unfortunately in every city the willingness to work far outstrips the opportunities now open. Most of us prefer to keep the land and the people idle.

The following is a sample of many letters received and shows the importance of this opportunity:

"New York, Sept. 29, '06.

Dear Sir: I have about one-half acre which my little ones and my wife have been working. Here is a list of some we have sold:

Radishes	$ 3.40	Peppers	.75
Celery	3.00	Egg Plant	.60
Parsley	.70	Pumpkins	2.40
Tomatoes	17.25	Turnips	1.50
Corn	5.10	Beets	2.00
Beans	2.30	Cabbage	2.15
Carrots	1.70	Potatoes	12.00
Cucumbers	3.75	Muskmelon	2.50
Lettuce	2.10	Kale	.75
Onions	4.15		
Peas	2.25	Total	$70.35

Besides this our cellar is filled for the winter, and I am satisfied, my six children and my wife are well provided for. Yours truly,

———————."

In addition to the amount sold and stored this man had supplied his table since June 1 with all the vegetables needed by his family besides taking in four boarders for several weeks. A low estimate of the vegetables used by his family and boarders for twenty weeks would be $100, and this must be added to the amount sold and stored to get his total production. This would amount to at least $200 in all.

Another remarkable showing was made by an old man. On a plot 60 x 100 feet he raised fifty dollars' worth of products, entirely by his own labor. At that rate of production he could have produced $360 worth of products to the acre.

Ninety gardeners started in at the beginning of the season and less than five per cent. failed to carry their work to completion.

Gaylord Wilshire, a prominent Socialist, and Editor of *Wilshire's Magazine,* says:

"In our grandfathers' days 'necessary machinery,' meant an axe, a hoe, and a log cabin, all of which were easy of individual production and ownership. To-day, 'necessary machinery' means a combined reaper and harvester, made by a one-hundred-million-dollar trust, a one-hundred-million-dollar railway to haul the wheat to market, a million-dollar elevator to unload it, and a million-dollar mill to grind it into flour, and finally a hundred-million-dollar trust to bake it into biscuits for all America." (Wilshire's Editorials, page 210.)

But the Vacant Lot Gardens show that even to-day available land only and not capital is necessary to make a living, and that any person who can get a bit of land can succeed upon it if he will work with his head as well as his hands.

CHAPTER IV

REASONABLE PROSPECTS

Living Costly—Hunger—Value of Food Products—What a Man Wants to Know—Fortune in an Acre—Stony Wold Record—Old Methods—New Methods—Acre Profits — Irrigated — Shearer's Success — O'Brien's — Hartman's—Small Gardens—A Woman's Patch—A 40 x 50 Garden—A City Backyard—Five Cents Per Square Foot—Glade Lands—An Illinois Plot—A Michigan Experiment—Farmers as Robbers—Youthful Gardeners—With Brains—Average Yields—Census Reports—What Averages Imply—Bailey's Estimate—Philadelphia Gardeners—Uncommon Vegetables—Other Callings Similar—Scientific Farming—The Farmer's Returns—What Could be Made—What He Makes.

"LET US HAVE THE FACTS."—Joseph H. Choate.

DUN & Co., the commercial agents, calculate that the cost of food has increased over one-half in the last ten years. Wages have advanced less than twenty per cent. Now they seem likely to recede.

In the winter of 1905, at the height of the boom, I saw a double line of men standing in the bitter wind at eleven o'clock at night waiting their turn for the cup of coffee that a newspaper gave—there was not one overcoat among them.

A look at these "bread lines" will convince anyone that even if "the amount of food that the world could consume is limited," as Dr. Engel thinks, we are still far from the limit, and that a fall in prices from increased production would be no unmixed evil.

These people are hungry because they can find no opportunity to work; one of the first steps toward their finding it is to show them that they need access to the land.

The food products of the United States in 1900 were worth $1,837,000,000, but the material that was grown for use in textile, leather and lumber industries alone was worth fourteen million dollars more than that. So we need not hesitate for economic reasons to send the city man to the Farm. Nor need the city land owner fear that he will suffer by the new farm movement draining away his tenants out of the city.

Increased production and wider prosperity will help his land values; for it is self-evident, when one thinks of it, that any improvement in the condition of the earth must go first and mainly to the owners of the earth.

But the hard pressed city man does not want theories nor statistics; he wants to know what show there is for him, and where he shall go and what he is to do when he gets there. He wants to know how he is to earn more than he is earning and more surely, by his own effort, and what land to buy and how to use land that may make him rich through the efforts of others if he only knows how to hold on to it.

He does not need much land for either purpose. An acre in the Bronx Borough of New York City, where I saw vegetables growing less than ten years ago, will sell to-day for more than a hundred thousand dollars.

"One man in one year, as I have understood," said Carlyle in "Sartor Resartus," "if you lend him earth, will feed himself and nine others." To-day one man, with access to an acre, can feed scores, no matter where that acre may be.

Now the Stony Wold Sanitarium for consumptive girls is at Lake Kushaqua in the Adirondacks, where two feet of snow fell on the 8th of April, 1907. But one man's work supplied over 150 persons with all the garden truck they could use, from May to November, and fed a lot to the chickens and cattle, off one and three-quarters acres. Besides that, as Dr. Goodall writes, they got forty-five bushels of potatoes and a large quantity of root crops also, to lay away for winter.

Taking those figures to pieces we find that, even in that climate, 1750 feet above the sea, where the snow lies into April and frosts always come in June, one-tenth of an acre will feed five persons and leave a surplus. The rest of an acre will give room for fruit or flowers to sell and will keep enough chickens and a litter of pigs to supply the most of the animal food desired.

Prince Kropotkin gives a higher estimate than this. In "The Conquest of Bread" he says: "Two and a half acres of market-garden yield enough vegetables and fruit to richly supply the table of 350 adults during the year. Thus 24 persons employed a whole year in cultivating 2 7-10 acres of land, and only working five hours a day, would produce sufficient vegetables and fruit for at least 500 individuals." But this can be done only by intensive methods.

A. R. Sennett in a recent publication entitled "Garden Cities in Theory and Practice" shows that it requires at least two acres of farm land, as at present cultivated, to feed each one of the people of America with grain and vegetable products.

Again, he estimates that it requires from two to three acres of cultivated pasture land to feed an ox, cow, or horse for a year, and, allowing one ox as the animal food sufficient for three persons per year, it requires at least an additional acre of pasture land on which to raise our beef or animal food; or three acres to feed each person. These estimates are based on the present ordinary wasteful methods of culture and pasturage.

The difference between such methods of pasturage and what has been accomplished by using

A LITTLE LAND AND A LIVING

intensively cultivated fields for permanent pasture, may be seen from the following table of the product of an acre in various crops.

	Average crop per acre in lbs.	Equivalent to feed in dry hay in lbs.	Number cattle fed yearly from every 10 acres
Unirrigated meadows and clover pastures cut twice per year	4,800	4,800	2.6-10
Swedish turnips cultivated	38,500	10,000	5.2-10
Rye, well cultivated	64,000	18,000	10.8-10
Sugar Beets	64,000	21,000	2.16-100
Indian Corn ensilage	120,000	30,000	3.30-100

In the first case two acres of land are required for each animal. In the latter, the two acres will feed six animals and leave six-tenths of enough to feed another. Merely a difference in what is raised.

By calling to our aid the latest scientific culture of food products we can show even a greater difference than in the feeding of cattle.

One irrigated acre has for thirty years given Samuel Cleeks of Orland, Glenn County, California, a larger net income and a better home than many of his neighbors get from hundreds of acres apiece. Mr. Cleeks saves an average of four hundred dollars per year after getting a good living from his acre, while many are becoming poor trying to run big farms without irrigation.

In the Eastern and Middle States are chances to do just as well on a single acre. Oliver R. Shearer of Hyde Park, near Reading, Pennsylvania, makes $1200 to $1500 a year on 3 1-3 acres, of which he cultivated 2 1-2 acres. He has raised and educated three children and paid $3800 for his property out of the profits of his intensive farming.

D. L. Hartman, of New Cumberland, Pa., in 1905, got $454 from an acre of early tomatoes and an equal amount from an acre and half of later tomatoes. An acre and a half of strawberries brought him $555 and his early cabbages averaged about $300 per acre.

He says that no one can fix the limit of value one acre can produce. One-sixth of an acre planted in radishes and lettuce, followed by eggplant and cauliflower, and the next year to radishes alone, followed by eggplant, brought over $200 each year; at the rate of over $1200 an acre.

A small plot 20 x 65 feet was planted first in pansies sold in bloom, then in radishes, part of which proved a worthless variety, then idle long enough to grow another radish crop, then half in late lettuce and the other half in winter cabbage which yielded no cash return. Yet $86.78 was received from this one thirty-second of an acre, at the rate of $2780 per acre. This amount could have been raised to $4000 an acre; all without using glass.

A woman on Long Island cultivated a patch of garden 25 x 50 feet and raised radishes, lettuce, onions, peas, string beans, carrots, beets, sweet corn, potatoes, tomatoes, lima beans, eggplant, peppers, parsnips, squash, and cucumbers, enough to feed her family of three and cleared $50 on sales in one season. This is at the rate, for the things sold, of $1750 per acre, after paying a man to spade the plot, for manure to fertilize, and $1.00 for seed. Even a city yard, though only 25 x 30 feet, if properly worked, can be made to produce enough vegetables for a small family.

A lot 40 x 50 feet, by careful cultivation,

yielded radishes, lettuce, onions, peas, beans, cabbage, beets, sweet corn, potatoes, cucumbers, sweet potatoes, and tomatoes, amounting to more than $20 in value, which was $433 per acre. Such returns are not confined to pieces of naturally "good" soil, but may reasonably be expected from any soil properly cultivated.*

In the mountain region of Garrett County, Md., there are swampy lands with streams running through them, known as glade-lands. A gentleman gave plots of these glade-lands to men employed in his tanneries, to be used as they pleased.

Truck gardening was the result. In the fall, when the crops were gathered, the gentleman had the yield of potatoes carefully measured twice, to be sure that the result was correct.

* The *Garden Magazine* for May, 1907, is the authority for an account of a city backyard garden 28 x 28 feet, which produced enough vegetables to supply a family of three from the middle of May until November. Successive sowings were made wherever a little space could be found, with the result that all the ground produced two crops, and most of it three during the season. The produce included radishes, lettuce, parsley, onions, strawberries, currants, peas, beans, salsify, tomatoes (early and late), corn, cucumbers, celery and winter squash. The value of the yield was $30.80.

The yield of potatoes was at the rate of 900 bushels per acre.*

Mr. E. A. Sutherland of the Nashville Agricultural and Normal Institute of Madison, Tennessee, writes: "I leased an eighth of an acre in Battle Creek, Michigan, and put it into ordinary garden vegetables. This little plot of land produced me green vegetables that would have cost me $80 (or $640 an acre) on the market. I kept a strict account at the time because I was desirous of knowing just what could be done. I took no other special pains with my garden beyond giv-

---

* Farmer's Bulletin No. 149 of the Department of Agriculture, in 1902 says: "In order to secure data regarding the amount of labor involved in the care of a garden, and the amount of produce it would yield, a 'farmer's garden' was planted at the horticulture department of the University of Illinois in 1901 so as to furnish a continuous supply of vegetables throughout the season. The garden was 280 feet long and 77 feet wide, or about half an acre. It was manured with 20 loads of well rotted manure, plowed early in the Spring and well worked down and then planted in long wide rows, so that most of the cultivation could be done with a horse. A succession of the same vegetable throughout the season was secured by planting early, medium, and late varieties, or by planting the same variety at different times. A combination of these two methods was found most satisfactory. The cost of all the seed used was $5.45. Putting a low estimate on the value of the crop raised, the vegetables could not ordinarily have been bought for $83.84. What other half acre on the farm would pay as well?

ing it ordinary care and cultivating it often. I was president of the Battle Creek College at the time, and was carrying heavy work, so could put only a little time each day in the garden. I did not sow another crop on any piece of the ground as soon as it was cleaned. I might have increased the value of the products considerably by raising two or three crops on a portion of it.

"Afterwards, I got about a third of an acre at Berrien Springs, Michigan, and cared for that for two years in about the same way. I had some fruit there. My own experience has been sufficient to satisfy me that an acre of land cultivated on the intensive plan will produce all the way from $300 to $1000. I find that this is a very common experience with those who can give the proper attention to the soil.

"I am convinced that there is very little systematic farming done in the United States. The most of the so-called farmers are simply robbers, who continually take from the soil without building it up. In a short time the soil becomes exhausted, and does not yield its strength. Then they doctor it with 'patent medicine' fertilizer.

"We have a tract of land near Nashville, upon which we are building an agricultural school. One of our first objects is to train men to take a piece of land and cultivate it in the right way, demonstrating what can be done with a few acres. We are making it possible for a man in the city to secure a home in the country and make a comfortable living."

To show what the least skilled labor may produce, the following samples are given: On a lot 10 x 10 there was grown in Philadelphia by school children ten to twelve years old:—

Beets,	6 bunches	.30
Cabbages,	3 heads	.15
Lettuce,	40 heads	$2.00
Lima Beans,	2½ pecks	.75
Radishes,	20 bunches	$1.00
String Beans,	1½ pints	.10
Tomatoes,	2½ pecks	$1.00
	Total	$5.30

This is at the rate of over $2000 an acre.

Mr. Edward Mahoney, Superintendent of the Garden School in Yonkers, N. Y., writes:—

"Our school covers one and three-quarters acres, the actual area cultivated by the children being 1 1-40 acres, the remainder being used for walks, observation gardens, tool house, flower beds, etc. The total value of the vegetables raised by the children last year was $1350.00. This value was computed from the prices the parents of the children paid to hucksters, stores, etc., for the same kind of vegetables. We use each year on our Garden School $150.00 worth of fertilizers.

Two students in the Colored Training School took a vacant lot and raised enough vegetables to supply two families all summer and sold enough to pay the taxes on the lot. A boy, fourteen years old, raised enough vegetables on an ordinary suburban lot to supply the needs of a family and sold $30 worth of truck besides.

Similar results are constantly coming to light from all over the country, proving that they are not confined to any specially favored section, but may properly be regarded as "reasonable prospects" from any gardener. It is a question purely of how much care and intelligence are given to the garden.

An old Master was asked, "With what do you mix your paints to produce such exquisite colors?" "I mix them with brains," he aptly re-

A LITTLE LAND AND A LIVING

plied. If we are to cultivate a garden or farm with our brains as well as our hands, small holdings should be selected as near to large cities as possible, so that large quantities of stable manure can be had cheaply. This not only greatly increases productiveness but also warms up the soil, thereby ensuring early vegetables.

If you don't *think,* you will make more money carrying a hod than you will cultivating an acre. It is not things like onions that require the most work, but things like blackberries, asparagus, which require the most intelligence, that pay the most. The following are average crops per year:

Beets, 300 to 400 bushels.
Cabbage, 8,000 heads.
Carrots, 200 to 30 bushels.
Horseradish, 2 to 5 tons (it sells for ten to fifty dollars a ton).
Onions, 300 to 400 bushels (but this can be doubled).
Potatoes, 75 to 300 bushels.
Rhubarb, 36,000 bunches.
Salsify, 200 to 300 bushels.

(These are actual averages per acre shown in Census Bulletin No. 237).

But in averages, the crop of the man who farms with his head and gets big results is "averaged" with fifty who use neither brains nor fertilizers. Like the school teacher who asked what was the average wealth of a class of twenty which had altogether twenty dollars, and was told one dollar. But when he asked if they were not prosperous, the boy at the foot said: "That depends on who has the twenty dollars." Yet good averages imply some wonderful yields.

In the "Horticulturist's Rule Book" Prof. L. H. Bailey gives the following table of average yields per acre in vegetables and fruits:

Beans (green or string)	200 to 300 bushels
Beans (lima)	75 to 100 bushels
Beets	400 to 700 bushels
Carrots	400 to 700 bushels
Cranberries	100 to 300 bushels
Cucumbers	150,000 fruits
Currants	100 bushels
Kohlrabi	500 to 1000 bushels
Onions (from seed)	300 to 800 bushels
Parsnips	500 to 800 bushels
Peas (in pod)	100 to 150 bushels
Potatoes	100 to 300 bushels
Salsify	200 to 300 bushels
Spinach	200 barrels
Tomatoes	8 to 16 tons
Turnips	600 to 1000 bushels
Apples (trees 25 to 30 yrs. old) alternate years	25 to 40 bushels
Peaches (in full bearing)	5 to 40 bushels
Plums	5 to 8 bushels
Pears (20 to 25 yrs. old)	25 to 45 bushels
Blackberries	1600 to 3200 quarts
Raspberries	1600 to 3200 quarts
Strawberries	2400 to 9600 quarts

A LITTLE LAND AND A LIVING

You've read many advertisements of fortunes to be made in tropical plantations, yet the yields advertised as marvellous are only $500 to $1200 per acre. Many market gardeners and nurserymen near our great cities can beat that. Don't take chances in a wilderness when you can do better at home.

Philadelphia market gardeners pay $25 to $50 an acre rent for five to forty acres each. Land as fertile can be bought for less than this, but they are right at the market and can sell their produce direct. Manure costs nothing and they can get the contents of privy wells delivered on their farms free; trolleys and telephones are at their service and they market two to three crops a year. They employ several men for each acre. They study to find, not what most people grow, but what crops will bring the most profits.

In his "Book of Vegetables and Garden Herbs" Allen French says that Jerusalem artichokes* are more profitable than potatoes and

---

* Farmer's Bulletin No. 92, page 21, 1899, tells of an experiment made by the Department of Agriculture in feeding pigs upon artichokes at the Oregon station. A portion of the plot was

just as valuable for food. They will yield from 600 to 1000 bushels per acre, will grow in any dry soil and may be used like potatoes.

The French Globe artichoke is the favorite for table use, and is so little grown here that it brings a good price in the market.

Seed of the French Globe artichoke can be sown in boxes in the house or in a hot bed about the first of March, and set out in the open ground about the end of May, in a deep, rich, moist soil, not too wet. Generally they do not bear until the second year, although if they get an early start and the season is favorable they may begin to appear in September. Like asparagus, they will bear for many years. In the North they need some winter protection; they should be tied to stakes in November, and the whole rows above the tops of the plants covered with earth and a layer of stable manure. The edible part is the flower head, which must be cut before fully ex-

---

measured and the artichokes dug to determine the yield, which was found to be 740 bushels per acre. The pigs gained 24 pounds from October 22nd to December 11th at a saving of two pounds of grain for each pound of live weight over the usual methods of feeding.

panded, boiled, well drained and served with Hollandaise sauce. No vegetable is more delicious.

Asparagus is a profitable crop. The main cost to establish a bed is labor, manure, and the use of the land two years. Part of a crop can be had the second year without injury. To maintain a bed costs per acre about:

```
For manure (applied in the Spring)..................... $25.00
Fertilizer (applied after cutting) ......................  15.00
Labor, plowing, cultivating, hoeing, etc..................  20.00
Cutting and bunching....................................  40.00
                                                         -------
                                                         $100.00
```

When well established, say in five years, and well cared for, it should produce 1800 to 2000 bunches a year for the next ten to fifteen years. At the factory price of ten cents a bunch this would be $180 to $200.

By free use of proper fertilizers and through careful culture, it is said that six tons per acre can be raised, which at $100 per ton would be $600, while the cost would hardly be $150 per acre, leaving $400 per acre profit.

Brussels Sprouts cost to grow $30, yield frequently over 8000 quarts of miniature cabbage heads per acre, which sell at 10 to 30 cents per quart. Average net returns $555 per acre, but the market is limited.

Cabbage average twenty-two tons per acre. Price from $8 to $10 per ton. Easy to grow, gather and pack. One grower netted $935 from three acres.

Cabbage seed, one of the specialties of Long Island, which is the biggest producer, nets over $400 per acre, but is not an easy crop to grow.

Cauliflower, where there is a good rainfall, is a delicacy that can be grown in large quantities in the open air. The crop requires care, but protected and blanched, its floweret-formed head nets a profit per acre averaging over two hundred dollars.

Other vegetables easily raised and profitable, although not common, include cardon and chard, of the artichoke family; sea-kale, leek, lentils, corn-salad, kohlrabi, celeriac, and egg-plant. But with all these you must be sure of your market in order to be likely to get a profit. Chard may be made by cutting back Globe artichoke leaves and tying the new growth of leaves at the tips to blanch for about a month. All these vegetables grow well in temperate latitude.

Tobacco pays on the right soil—2000 pounds per acre can be raised. Connecticut tobacco brings twenty to thirty cents a pound, or four to six hundred dollars per acre for a good crop. Some soils in Connecticut raise tobacco equal to Sumatra, which sells here for about $3.50 a pound. It is claimed that in parts of Pennsylvania and Ohio Cuban tobacco can be grown. Soil is of such importance in tobacco growing that the Department of Agriculture is making soil maps of the important tobacco sections. Write for them if interested.

Mushrooms can be grown as a supplement that interferes with nothing else in the winter time. The new method of germ isolation and spore culture has cut out much of the uncertainty. The only things needed now, besides knowledge, to insure success is an even temperature, careful attention, plenty of good barnyard manure and commercial sense enough to get your product sold to advantage. An outhouse, cellar, limekiln or waste space for a greenhouse will answer for mushroom culture.

Don't forget to find out about the wild mushrooms that are good eating, for your own benefit. There is only one poisonous family, whilst there are more than a hundred wholesome ones that commonly grow wild. So it is only necessary to learn the dangerous ones and then eat with confidence any mushroom that tastes nice, provided it has pink gills or looks like coral or grows on wood or is a tender young puff ball.

That there is another side to agricultural prospects the following letter states clearly:

"Those who do any farming are rarely very enthusiastic about it. There is something truly diabolical about the uncertainties of nature in connection with growing things, when it is necessary to get definite results. In chemistry it is possible, by experiment, to get exact results. We know that certain things, in definite amounts, will always produce a definite something. We can be absolutely certain. But in farming it is not so. Man can study his soils, and plant his seeds, ever so scientifically, but there is an element that controls the results far more than seed or soil, over which man has no control, and that is the climate. Nature performs such freaks in this domain that farming seems a little less certain than most forms of gambling. For instance, a cold north wind has been blowing down here in Alabama, for the past ten days or more, that is blasting everything—after they had been growing for weeks in the soft summer air. Once, in Colorado, we had ten acres of navy beans. They were growing wonderfully; we expected to make between $1000 and $2000 easily; had the beans all sold to hotels in Denver, before the crop was half grown. It was the marvel of the whole country; people came from far and near to see those beans. We were not farmers; had no experience. Just pure "luck." We were spending the summer out there, and raised the beans to see what we could do. Well, *early* in August there came a hard freeze one night. The next morning our beans were utterly destroyed. The "oldest inhabitant" had never heard of such a freeze that time of the year. We had a neighbor in Colorado who in the early days made $900 one year on potatoes. He has farmed ever since, he and his whole family working like slaves, and they barely make a living. Who can tell what made such a truly miraculous yield that one year? It is beyond the subtlest science, evidently. I know dozens of cases like these. Men who devote themselves to farming for years, who study every detail with the greatest care, know that they never can tell, an hour ahead of the actual harvesting of things, what they will get. I wish it were not so. It makes living a hard proposition, even if the land was properly distributed, and all had a chance."

This is all true of farming big acreage for little crops. But before the chart and the compass and steam came into use, the sailor had just such risks as our farmer has now. He also was at the mercy of the weather.

Now, what are the risks of the farmer? Tornadoes? A lumber business or a factory is equally exposed to those. Drought or excessive rains? Cultivation or irrigation protect against the one, and trenching and draining against the other. Insects and diseases? We know how to combat those. Late frosts or early frosts? Hot beds and setting out plants avoid the late chills and bring the vegetables to maturity before the early frosts. There remain summer hail storms, but compared with fire and other risks in other businesses, this is insignificant.

If that farmer, instead of working like a slave, had thought like a scientist, he would know why he had made big profits one year and only a living since.

In fact, the scientific farmer, who cultivates not scores of acres, but the little patch that he can watch and manage, is in no more danger than any other business man, and he does not suffer from strikes and cannot be discharged.

But one may not ignore the condition of the ordinary farmer. Let us take the actual case of a farmer living near Ghent owning 160 acres, of the average value of $20 per acre; total $3200. One hundred and fifteen acres are in cultivation. If this produces $87.50 per acre, the gross average in the New York Vacant Lots, we have a gross value of $10,062.50. Now, the New York "Vacant Lots Cultivation Report" based its estimated valuation on wholesale market prices, but retail prices were generally received by the planters. The U. S. Agricultural Department calculates that the farmer receives 60 per cent. of the retail price. This gives him gross, $6037.50. That is what he ought to get.

Out of this he has to pay interest on mortgage of say $1300, $78; taxes, $48; seed, $60, leaving $5851.50 for wages, machinery, tools, food, clothing, insurance, repairs and general expenses.

On this farm it takes two young men to do the work, so we must divide all the items and the result in halves to represent the result of the work of one. We find $2921.25 as his gross income, after paying the interest, taxes and seed.

On this showing he would do very well; but see: The farmer whose real expenditures we have taken is forced to till poor land, and tills far too much of it, and probably, cut off by his isolation from the sources and stimulus of instruction, he tills it badly, and the actual product is only $264.50 for each man, less $93 each for the three items specified, leaving $171.50 each for all other expenses and for " profit." That is what he does get.

A LITTLE LAND AND A LIVING

This is a very different result from that of the poor people, working land near New York on an intensive scale with good instruction.

Of course in most cases, seed, and in some cases, fertilizer, were given these people, but we must remember that if they had even a five years' lease of this unused land they could supply their own seed and pay this back besides, and that if they had the use of all the fertile unused land about our cities they would not need to buy fertilizers.

CHAPTER V

RECORD YIELDS

Possibilities—Production and Cost—A Standard—Records—A Garden for Five—School Gardens—Small Plots—Yields of "Poor" Soil—Celery—Texas Onions—Corn Record — Strawberry Yields — Rhubarb — Christmas Trade—A House-cellar Patch—Forcing Cellar—Onions by New Culture—Asparagus.

NONE of the figures we give show the full possibilities of an acre. No man can measure that; the only question is how far it will pay to increase its products. You could not hope to "beat the record" with a horse or with a yacht unless you have hundreds of thousands of dollars to spend. Nor to make a new record in the mile run or the high jump without a special physique or long training; but by experiment through tillage and care, an ordinary person may surpass the best records shown in crops, or may develop new trees and plants that will benefit the world. Hitherto, truck farmers and market gardeners have originated most modern improvements in farming.

The farmer does not usually even notice how much he raises per acre; he has plenty of land, and the only question that interests him is how many bushels he can get altogether and what they cost. Besides, he has neither the time nor the education to tabulate results.

The Experiment Stations are so occupied with more pressing problems—insects, varieties, hybridization, inoculation, fertilizers, and so on, that they have not given much attention to the maximum increase of crops nor to the proportion of increase of cost.

A healthy interest could be excited and much valuable information be obtained if the Agricultural Societies and Granges would ask the experiment stations and progressive farmers to figure out records as standards for each crop raised in their neighborhood by which progress could be measured. We have abundant well established "records" of high jumping, of speed of horses, of automobile runs and a thousand other things, why not of the productivity of acres?

While many elements enter into this problem, only a little common sense and practical knowledge are needed to fix profitable standards, and it would bring great benefits. A standard would differ materially from an average. All averages are misleading, especially those shown by the census, and unless you use brains and raise more than the average, you will make but a bare living, though comfort and even luxury are easily within your reach.

It would be easy to get a lot of unverified newspaper clippings about astonishing harvests, and profits like the ghost stories in the mining advertisements. But the records given here have been investigated by the writers or else rest upon unimpeached authority, which we quote. Unquestionably they are not the best ever, and possibly some errors will remain; but they are given in the belief that they will bring out many better records, and perhaps some corrections.

However, the record that interests you the most, is the record of what you can get and how.

The old way of getting a bare living from much land and great labor has no attraction for the average man, nor should it have. If Nature were the niggard she has sometimes seemed to be, we would owe her no gratitude. In reality she has provided a wholesome and bountiful living for the average family in every acre of soil, and if we do not get it, the fault is our own.

It must be borne in mind that in most cases three or four crops or more can be grown on the same land during one season.

If we raise only one crop, we will not get a good living from even ten acres; that is no fault of the soil; it is due rather to ignorance or carelessness.

A few years ago Professor Bailey said that 100 x 150 feet of land, if well tilled, will give a family of five an ample supply of all kinds of vegetables, except potatoes.

Now comes Mrs. Helena R. Ely and says that a plot of ground 20 x 80 feet, or 1-5th of Prof. Bailey's estimate, if well fertilized and well cared for, will yield cauliflower, egg-plants, lettuce, peppers, parsley and tomatoes for a family of eight persons. If one chose to vary the list somewhat the results would be about the same. Two five-cent packages of each variety of seed, except cauliflower, which costs ten cents, will raise more than sufficient plants. The seeds should be sown in March, either in boxes in a sunny window, or in a hot bed, and set out about May 20th; or plants may be bought and set out at that time. A very little regular attention is all that is needed by such a plot.

By putting in a new crop as soon as one was harvested, School Garden boys, eleven to twelve years old, raised on a sixteenth of an acre, 336 bunches of radishes, 110 bunches of onions, 368 heads of lettuce, 89 bunches of beets, 8 bushels of beans, 7 bushels of tomatoes, 7 bunches of carrots, and 1 peck of turnips, besides nasturtiums and petunias, many boxes of which found their way to the hospitals of the city. This was at Washington on the grounds of the Department of Agriculture. At regular market prices $55 worth of produce was gathered from the small plot. This is at the rate of $960 per acre. The report naively remarks, "experienced farmers sometimes fail to do as well." (U. S. Bulletin No. 160.)

A LITTLE LAND AND A LIVING

The best lay-out for an acre garden for easy cultivation is about 132 x 330 feet, and easy cultivation is an important factor in creating profits. In one such garden in Pennsylvania, one-half acre was planted with strawberries, one-half with early beans, beets, cabbage, cauliflower, celery, cucumbers, lettuce, melons, onions, peas, potatoes, sweet corn, tomatoes, while the last row was set with Hathaway raspberries and Blowers blackberries. This garden fed a large family, provided vegetables and fruit for winter canning, and furnished $300 worth of surplus vegetables and fruit for market.

On Prof. W. G. Johnson's truck farm in Palatka, Fla., between the sowing of lettuce seed on September 23d and the first of January following, the profits of half an acre amounted to $295. Three crops of lettuce were gathered from the same land during the year, with celery and cucumbers intervening. The celery yielded from 650 to 850 crates per acre, ranging in price from $1.25 to $3.00 per crate.

Half an acre of land in Bath Co., Kentucky, on the farm of V. C. Razor, was made productive by successive crops. In 1905 it was planted first in strawberries, which sold for $60.00. The second crop was of sugar corn and pole beans. Besides supplying two families, $12.00 worth of corn was sold, and the poor ears were used to fatten a pig. Of the beans, some were sold, some given away, a great many were eaten by the two families, and a bushel was dried for winter use. When the frost came, rye was sowed and harrowed with a disc harrow. This furnished excellent pasture for pigs and chickens all winter, and in the spring was turned into the ground as manure.

A 25 x 25 foot garden at Rhode Island Agricultural College yielded vegetables worth $32.18 at market rates, or at the rate of five cents per square foot per acre. Three cents per square foot is considered a good gross yield.*

* One of the students at the Normal School at Washington, D. C., cultivated half of his back yard, and raised enough vegetables to feed a family of eight persons all summer. This garden patch was about 18 x 15 feet. At that rate an acre would have fed 1290 persons.

A preacher in Indiana Co., Pa., on a patch equalling about one-fiftieth of an acre, raised beans, peas, some strawberry plants, and a fine crop of tomatoes. The vegetables supplied all the wants of a family of four.

To prove that any soil could be made productive by reasonable care and preparation, Mr. C. E. Conwell, a business man of New York City, chose two acres of poor, hard, clayey soil, almost impossible to plow, for potatoes. The land was plowed in June and sown with a mixture of oats, rye and peas. In August this whole crop was plowed under and then harrowed in. The land was next plowed in May and planted with Peerless potatoes on May 9th. The yield in October was 964 bushels of perfect tubers which sold for 50 cents a bushel. Mr. Conwell says that his expense for that first yield was $156.15; his returns $482.00, leaving a balance of $325.85 on his two years' work. The following year the same plot yielded 985 bushels. After that, for several succeeding years, the yield fell off slightly, but what was lost in quantity was more than offset by improvement in quality.

But large returns are not restricted to potatoes. Alfred P. Edge, of Hartford Co., Maryland, set out celery plants in rows 6 inches apart and 3 inches apart in the rows, in a bed 54 x 6½ feet, getting 11 rows of 230 plants each, or a total of

2530 plants. He sold and used 2220 plants, receiving 5 cents each for 720, and 4 cents each for 1500, making an income of $96.00 from 352 square feet.

At this rate of yield and price an acre would give 277,500 plants and bring in $12,000. But it would not be practical to plant an acre so closely, because it would make cultivation almost impossible.

Kalamazoo, Mich., is the real centre of celery culture in America, about 200 acres of marsh land being under cultivation there. Most of the work is done by Holland families, the women and children working in the fields with the men. One man can tend two or three acres, or from 40,000 to 60,000 plants. The returns average about $200 to the acre.

Before the severe freeze of 1895 oranges were almost the only crop grown in Orange County, Florida. Celery growing has since developed and is becoming more general every year. Twelve hundred crates to the acre is considered a good yield, the income therefrom being $1680. This is net income at point of shipment.

The possibilities of onions as profit makers are only now becoming known.

Referring to his success in raising Bermuda onions on his farm in Texas, Mr. E. L. Hoffman writes:

"The statement in the land advertisement that we cleared $10,000 net profit from 20 acres of Bermuda onions is not exactly correct. We, however, did clear a net profit of $500, after paying freight and commission, from one single acre of onions, but the whole 20 acres did not average us quite $300 an acre. In rough figures I could say that last year's crop ran about 18,000 lbs. to the acre."

Mr. John Closner, of Hidalgo, Texas, says he made a net profit of $11,000 on 33 acres of onions. The yield was 36 car-loads.

The average onion yield in the United States is 200 bushels per acre, although in Connecticut it runs from 350 to 500 bushels. E. N. Foote, of Northampton, Mass., has made a specialty of onion-raising and this year has averaged 800 bushels per acre, with some acres yielding 1000 bushels. Formerly his highest yield has been 910 bushels, but he expects to get 1200 from some of his acres later on. His crop this year, covering 25 acres, brought from $12,000 to $14,000.

Corn is the staple American product, but there are many other things that give larger returns for time and labor expended. The yield of corn per acre for 1906 was 31 bushels; this is the general average. The prize offered by the *American Agriculturist* for the best yield for that year was won by J. A. Tindale, Clarendon County, S. C., with 182 bushels. But this does not touch the world's record, won in a similar contest in 1889 by Z. J. Drake, of Marlboro Co., with 217 bushels of chemically dry corn, or 255 bushels green weight, to one acre. That little 150 bushels of corn extra which Mr. Tindale raised was the reward of intelligent effort; but unless someone will give you a prize it won't pay to try for it.

Henry Jerolaman, of New Jersey, is known as the Strawberry King of the World. His farm has been producing strawberries for more than forty years. When it was the property of Seth Boyden, it produced in 1869 the world's record for size, and the berry was called after the

American Agriculturist. Mr. Jerolaman has beaten that record more than once; Prof. W. G. Johnson says that Mr. Jerolaman has produced berries four inches in diameter. Four inches diameter would make a big apple. They hold the world's record and are likely to for some time to come.

Large, sweet, good flavored berries, Mr. Jerolaman says, sell on sight, and never glut the market. Of the berries raised on his farm, thirteen usually fill a quart box; he considers 6,000 quarts per acre a good yield.

Mr. Frederick Wright, of New Jersey, is authority for the statement that the Climax Strawberry has yielded 6300 quarts on less than ¼ acre. Mr. Wright saw the field in bearing in the ground of the originator of the Climax, Mr. H. W. Graham, White Haven, Md. (in 1903, I think).

Two years ago, Mr. T. C. Kevitt, Athenia, N. J., had an acre of Glen Mary strawberries that yielded 21,780 quarts. Mr. H. W. Collingwood, Editor of *The Rural New Yorker,* went over to Mr. Kevitt's, measured the acre, and saw a portion of the berries picked, from a measured space, and was convinced of the yield. Mr. Charles Wiley writes: "On a small plot in Bay Shore, on the place on Park Avenue which I sold to Mr. Cartwright, I raised at the rate of 85,000 quarts to the acre of strawberries." That would carpet over two-thirds of the acre with a layer of good-sized berries.

Perhaps the most wonderful story of strawberries is that published in the *Rural New Yorker,* April 20th, 1907, and supported by affidavits. Samuel Cooper, of Delevan, N. Y., grows strawberries out of doors as late as when the first snow falls. During 1903-'04,-'05,-'06 he supplied late strawberries from August 7th to the end of October, to the Hotel Broezel, Buffalo, N. Y., as well as to local customers.

The largest quantity shipped on any one day in October was 20 quarts. Soon there will be no season without this luscious fruit.

But strawberries have no monopoly as revenue producers. Any garden product that is better than the average of its kind, will fetch a good price, particularly if "out of season." Under old methods each season brought its own products; now we have almost annihilated seasons so far as garden truck is concerned.

Mrs. P. Bailey, of California, says: "When a girl, I had the picking and selling of the strawberries, and I see by my old note book that I sold more than $100.00 worth of berries from our bed. It was about 20 x 20 feet, if I remember rightly, and during the month of June I sold more than $40.00 worth and had regular customers for the berries all summer."

That was at Stockton, California, but Mrs. Bailey now has a little home of her own at Salinas, and of this she says: "I get all the berries we can eat from a little bed of six rows—12 feet long, and have put up some and given away a few. There was a man called on Sunday and he was surprised at our little home garden and said he liked the strawberries best of all. As he has just bought a little home of his own, he wants a good patch of strawberries right away."

Rhubarb is another profitable crop, and J. E. Morse, in "New Rhubarb Culture," tells how it is forced for the Christmas market. All that is necessary is the complete exclusion of light and frost, maintaining a small degree of heat, and a little regular attention.

Although specially constructed forcing cellars are desirable for large crops, Mr. Morse tells of a rhubarb patch, 36 x 54 feet, in a house cellar which gave $144 in returns. It was partitioned from the rest of the cellar, kept perfectly dark and heated by two ordinary kerosene house lamps with chimneys smoked to prevent them lighting the patch. The light from even a small lamp will bleach the stalks near it. It requires little fertilizer, heat or moisture, so that there is no disagreeable odor in the cellar where grown, if the roots are taken up as soon as the forcing season is over.*

As forced rhubarb requires little soil, even a cellar with a cement floor can be used, two or three inches of earth on the surface being sufficient. Rhubarb is a hardy northern native, and does not thrive in warm countries. The best results are obtained if the roots are frozen before

* The expense of building forcing-cellars varies with their size, but need not be great, especially if old lumber is used. From such a cellar, 12 x 50 feet, the returns for one winter amounted to more than $160.00.

transplanting. The same roots will not do for forcing two seasons in succession.

The new method of cultivating vegetables in the field during winter has been successfully applied to rhubarb. Exhaust or waste steam is forced through numerous tunnels running between the beds, thus keeping the ground moist and at the proper temperature. This is preferable to heating by steam pipes. The returns from rhubarb thus forced are from $100.00 to $500.00 per acre.

As Greiner says in his "New Onion Culture," these new methods, accompanied by transplanting, are wonderfully successful when applied to onions. By the new culture an acre will yield from 1000 to 1500 bushels of onions. It cost $300 to produce 1000 bushels upon one acre; the price received was $750.00, leaving a profit of $450.00 for the acre.

But a prominent grower estimates that it costs only $100.00 per acre to prepare the soil, start seedlings, transplant, cultivate, weed and pull the crop when the new culture methods are followed. From 800 to 1000 bushels per acre is a frequent yield and will cost little more to produce than 500 bushels. Transplanting increases the quantity and greatly improves the quality of the yield.*

The new methods of culture apply to asparagus also, a crop not so generally grown as it should be, the prevailing idea being that it does not greatly reward attention. F. M. Hexamer in "Asparagus" shows what a mistake this is. Nothing better repays care, but it is necessary to grow it from good seed. One pound of seed will produce 10,000 plants, about half of which will be vigorous, the others should be thinned out. It requires about three years to get a thoroughly profitable crop.

A New Jersey grower cut 22,584 bunches of asparagus from 12 acres, all of which were not in full bearing, or 1882 bunches per acre. His net returns *from the commission houses* were $2611, or a little more than 11 cents a bunch. Higher

---

* The Department of Agriculture (Bulletin No. 39 on Onion Culture), says that at the Ohio station 10 selected transplanted Prize-taker bulbs weighed 8 pounds and 4 ounces. A decided increase was noted in the 14 different varieties tried, amounting in some cases to 100 per cent.

A LITTLE LAND AND A LIVING

prices and larger yields are common enough, but this is a fair average.

The average annual yield of one acre was 2500 bunches; value 12½ to 25 cents per bunch; the net yearly returns for ten years averaged over $550 per acre.*

Asparagus yields for thirty years, but good business policy dictates renewal after ten years' cropping.

---

* Great stories are being told about the profits of growing asparagus in California. We read of farms with over 100 acres of asparagus in one field, and also of profits of $1,500 on one acre. A Massachusetts asparagus grower who is ranked as a large operator here, sends a clipping and writes:

"I enclose one which tells of a profit of $1,000 per acre. Now, this is for blanched asparagus and is a long way ahead of anything we can show. Incidentally the white asparagus is not fit to eat, but if they get such 'profits' they do not care much for the quality. In New Jersey one man writes me of $1,500 profit on three acres exclusive of freight and commission. My best yield on a 15-acre field was $450 per acre. One of my townsmen received last year $1,000 from 1¾ acres, and another $450 per acre, but the rust had not hurt them much. Who under the sun can tell a 'bigger story' than a California westerner? Even Eastern people who go there catch the 'fever.' My wife came home from Los Angeles a few years ago and told me that some of the guests in a large hotel in Californa had to walk half a mile to their meals in the building! It reminded me of a Nova Scotia man who said they had trees so tall down there that it took a man and a small boy to look to the top of them. But really large stories are true in California."

Rural New Yorker, May 25, '07.

CHAPTER VI

WAYS OF WORKING

The Home Garden—First—Working the Soil—A Beginner's Garden—Save Labor—Fruit—Planting and Transplanting—One Crop Risky—Companion Cropping—Specialties—Marketing—Succession Crops—Deepening Cultivation—Subsoiling—Soil Enriches Itself—Insects Smothered by Ashes—Winter Plowing to Kill—Spraying—Fertilizers—Irrigation—Drainage—Dry Farming—Fancy Packing.

NO matter what the profits of the farm are to come from, the home garden should have first attention; that is what men live by, regardless of prices or hard times. It must be large enough to supply the family with food in abundance and variety, but, to economize labor, it should be larger than that. A quarter acre will give your family a succession of vegetables in summer and fall, and potatoes and turnips for the winter. It will take about four days to dig it over and four to plant.

The part kept for radishes, lettuce, beets,

spinach, Swiss chard, peas, string and wax beans, may be dug over for three successive plantings, but that used for early potatoes would need digging only twice; the second time when you are planting late cabbage or turnips. Plow it if possible, in the fall, so that the freezing and thawing will break up sods and clods. Before plowing, manure with twenty-five heaping wagonloads to the acre.

After manuring and plowing use a disc or cutaway harrow till the soil is fine as dust. Then your seeds will have a chance to grow. Mrs. Fullerton says: "Weeds are the farmer's best friends, they force him to cultivate, but every stone, weed or clod left, destroys part of the food your plants must live on. Don't be deluded with the sluggard's reason that "the stones warm the ground."

For beginners who want fresh fruits and vegetables from May till Christmas 100 x 200 feet is enough; a regular fruit garden alone takes about 100 x 100 feet.

To get the best results, the chief factors are nearness to market, rotation of crops, saving of labor in cultivation and keeping up the fertility of the land. When the land is near the market the product can be peddled or sold to customers, which makes a big saving of capital in crates, storage room and teaming.

To save labor we must plant in long rows each just wide enough apart to allow the furrows of the wheel-hoe to cover the whole ground between both small plants and large ones without readjusting the blades.

If you plant in beds you can't use the wheel-hoe well, and you'll probably plant yourself in your own bed with backache. Choose a southern exposure; as this gets all the sun, it will make the earliest garden. Lay out a plan of your land, and whether for home use or for market:—

(1) Plant in rows; 100 feet of anything is usually enough.

(2) Put asparagus, rhubarb, sweet herbs, and other permanent vegetables in a row at one side so that you can plow the rest easily. Run the rows north and south so that each row will get the sun from the east in the morning and from the west in the afternoon.

(3) Plant together vegetables of the same height, tall ones back at the north end so as not to shade the others; and see that there are woods, a hedge or a building for

A LITTLE LAND AND A LIVING

a breakwind against the prevailing winter wind; this will often give vegetables a fortnight earlier in the spring and from two to six weeks later in the fall.

(4) Plant vegetables that ripen at the same time near one another.

(5) Practice rotation; for instance, lima beans should not follow green beans or peas; and, as far as possible, put the plants subject to the same diseases and insects together.

A row of apple trees about fifteen feet apart might go on the northern border, plums and pears on the west, and cherries and peaches on the east.

Put a grape trellis next the apples, then a row of blackberries, raspberries, gooseberries and currants. These will form the windbreak to protect the vegetables besides adding income. Small fruits planted this year will bear next year. Then between these and the vegetable rows you can put the rhubarb and asparagus.

When it comes to planting it is important to plant seeds at the proper depth. If you plant them too thickly it is easy to thin them out when they grow.

Transplanting can be done on a large scale. Many vegetables are greatly benefited by it; it gives each plant room to develop, as well as economizing space. This is especially true of lettuce and salad plants. Lettuce may be started under cover in old berry boxes and the boxes then plunged in the soil.

Onions also gain greatly and about as much labor is saved in thinning and weeding as it takes to transplant. For fine grades, these gains are important. The soil must be kept busy all the time; if idle even for a week, weeds will grow which steal the food from your paying crops.

Ernest Rollenbeck says that as soon as early peas are harvested he plants squash. If sweet peas are grown, a row of onions may be grown on each side of the peas, without detriment to either. Late cabbage may be set in the rows of early onions and make their growth after the onion harvest. This gives three crops.

To raise but one crop is risky; its failure may ruin you. With a number of crops, what hurts one may help another, and even a complete failure of one will not be so serious.

Companion cropping grows two crops in the same soil at the same time, one maturing early and leaving the ground free for the main crop.

In this way late celery is planted between the rows of early celery; radishes with beets or carrots; before the beets need all the room the radishes mature; corn with squashes, citron, pumpkin or beans; horseradish or early onions with cauliflower or cabbage. Lettuce with early cabbage, etc.*

No one need fear that his land will become too rich.†

Specialties often pay better than general crops. The returns can be made immediate and the work almost continuous, through companion cropping.

Marketing is an important item in success,

---

* A complete planting table for vegetables is printed in the appendix of "Three Acres and Liberty."

† At a recent meeting of the New York Florists' Club, M. H. Weezenaar, of Holland, made the following statement as reported in "The Florists' Exchange":

"Land in Holland adapted for the cultivation of hyacinths, he said, runs from $2,000 to $3,000 per acre. The manure used amounted to something like $1,000 an acre, in addition to which there was a big water tax, levied for service of the pumping engines, the water having to be pumped at certain seasons from the canal.

The soil in which the hyacinth bulbs are planted is dug to a depth of three feet, the work being done in winter, the aim being to get as much frost as possible into the soil. When spring arrives, from 10 to 15 inches of pure cow manure is dug into the soil to and judging from the prosperous condition of the savings banks in Long Island and Jersey it seems clear that the farmers have studied the markets to good purpose.

One of the earliest centres for truck farming was along the Chesapeake Bay, where oyster boats were employed to send the produce to the markets of Baltimore and Philadelphia; so the gardeners about New York early began pushing out along Long Island, using the Sound for transportation. The eastern shore of Lake Michigan is another sample of the effect of convenient water transportation in causing an early development of this industry.

The time spent in " pro-ducing " is not only the time between putting the seed in the ground and putting the crop in the basket. The producer is one who brings things forward, and

a depth of one foot. In that soil are planted potatoes, peas or beans, either for seed or market. In August the products are gathered. The same soil is then dug to a depth of 1½ feet, cow manure being again applied to it. The beds are rounded and there is a ditch of about one foot broad and deep, dug along the sides of the beds to carry off the water. The bulbs are planted from 2 to 2½ inches deep. The beds are then covered with reeds to a depth of 10 inches. This covering is not put on until the severe weather is past, the object being to allow freezing.

production consists just as truly in getting it to market and seeing that it gets a fair chance there, as it does in buying the seed or in setting out the sprouted potatoes.* How much time may be spent in that part of production, depends upon the individual and it is not reckoned here. We calculate only the time spent in physical work, not the time spent in thinking how to do the work, which is more essential than money. You must be willing to work and work hard with your hands when necessary, but you can, after you get started, hire men who will not or cannot use their heads, to do most of the hard work. Corn, which everyone can raise, sells for half a cent a pound; mushrooms need care and skill and bring fifty cents a pound.

Vegetables may be grown in rows between the trees in a young orchard. The best beans grow in orchards. Radishes, lettuce and cabbage may all grow at the same time in space used for one crop.

* R. B. Greig in the Aberdeen and North of Scotland College of Agriculture Bulletin (No. 3) says that experiments with sprouted seed potatoes gave a gain of from one ton (say 35 bushels) to three tons per acre.

Plant corn after early lettuce and radishes are gathered, and more lettuce when beans are picked; then cabbage, cauliflower or spinach where the early corn grew, so that the small patch may earn your living and pay big dividends. Early potatoes and early cauliflower are followed by celery and Brussels sprouts.

Onion seeds are sown between early planted onion sets. Then cauliflower is planted. Later we may put a few cucumber seeds between the cauliflower. The onion sets mature first and are gathered, then the cauliflower in time to permit free growth to the cucumbers. Between the seeds and onions we may plant radishes, lettuce, peppergrass or spinach, which will mature before their shade could hurt the seedling onions. Later, turnips may be sown between the onion rows.

This gives two crops of onions as well as cauliflower, cucumbers, radishes, and turnips from the same plot, and it leaves no chance at all for weeds.

One can live while waiting for the crops to come up, for many crops mature rapidly. The accompanying diagram, taken from the United

A LITTLE LAND AND A LIVING

States Agricultural Bulletin No. 149, will show the time various crops ripen and how they may follow one another:

VEGETABLES.	MAY	JUNE	JULY	AUG.	SEPT.	OCT.
Radishes	—	—				
Cress	—	—				
Green Onions	—	—	—			
Lettuce	—	—				
Spinach		—				
Beet greens		—				
Peas		—	—			
Early beets		—	—			
Early potatoes		—	—			
String beans		—			—	—
Cabbage			—	—	—	—
Early carrots			—	—		
Sweet corn			—	—		
Tomatoes				—	—	—
Lima beans				—		—
Peppers				—	—	—
Parsley				—	—	—
Summer squash				—		
Cucumbers				—	—	
Muskmelons				—	—	—
Watermelons				—	—	—

See how one crop goes off to make room for another; most of them will be planted in the Spring and will be growing all together, but they don't need to take up your valuable land all together. You start the seeds in plates with wet brown paper, and the plants in "flats," shallow boxes of earth, or in window boxes, cold frames, hot beds or glass houses and set them out intelligently. How to do that is a study by itself. For it is necessary that the crops should get their growth at sufficient intervals not to steal nourishment from one another.

Of course, many crops, radishes for instance, may be grown, by successive plantings, right through the season, and late sweet corn may be planted as late as August 15th in the Central States.

Three acres will give a good living on this basis.

To produce big crops deepening cultivation is necessary. This is because the more soil surface exposed to the air, the greater the regain by that soil of the productiveness lost in former cropping. The market gardener who breaks and subsoils his land deeper each year down to a depth of 24 to 30 inches, will, *if the under soil is good,* greatly surpass the gardener who only plows his land to a depth of 8 or 10 inches. The subsoiling should be done in the early fall or early spring.

During drought the land should be thoroughly watered once a week and every precaution taken by careful drainage to prevent excessive wet. If cattle are kept, the soiling system should be pursued.

If you have land enough you can keep one or more cows, if you can find folk willing to pay an extra price for milk they know to be pure and fresh and from well kept cows. At times growers have to throw produce away. Sometimes it occurs through failure to find out in advance what vegetables are wanted and the probable extent of the demand. Such surpluses would help the milk yield very much.

If you can get land that has been in red clover, alfalfa, soy beans, or cow peas, for years, so much the better. Bacteria on their roots draw nitrogen from the air which becomes fixed in the soil. Nitrogen is the great meat maker and forces a prolonged and rapid growth.

Mr. A. Crosby, of the Bureau of Plant Industry in Alabama and Mississippi, says that a farmer who had red hill land increased the yield of cotton six fold, from one-third of a bale to two bales, by merely putting in a crop of bur clover in the winter. The clover reseeded itself. Not bad use of an acre, especially if the cotton follows lettuce, radishes or tomatoes. Two hundred dollars' worth of cabbage, two bales of cotton, say at $75 per bale, and $75 worth of turnips seems to be about the record for this sort of farming, but it can easily be beaten. (See Year Book —United States Department of Agriculture, 1905.)

Many farmers have found that the inoculation of the soil with nitrogen bacteria * or the growing of clover, soy beans, or cow peas, will make a considerable difference in their yields.

To manage insects (or infants) you must be interested in them and study them. It is fun and it pays in both cases.

Ashes slacked in lime, or any other dust or powder freely put on leaves when dry, will

* NOTE. Send to the Agricultural Department at Washington for the latest reports on this subject, which is an important and interesting experiment.

smother most of the insects so destructive to plants. Winter plowing kills cut worms. As each family of vegetables has its own peculiar bugs, constant change to new soil keeps the bugs from getting a good start. This is only one of the advantages of the rotation of crops.

Even if you raise fruits and nuts that need no hands and knees work (which in any case can be reduced to the minimum by the use of the wheel-hoe), the time to fence off borers with a circle of tin pushed a few inches into the soft soil is just when the borers threaten; the time to spray is before and after the bloom, and generally if you won't or don't do these things, they will be done too late or will not get done at all.

You would better run an office without a type-writing machine, than a farm without a spray pump. Spraying at the proper time may mean all the difference between profit and loss. "In fourteen co-operative experiments, covering 180 acres, made in 1904, in New York State, the average increase in yield of potatoes due to spraying was $62\frac{1}{2}$ bushels per acre, the cost of spraying was $4.98 per acre, the cost per acre for each spraying, 93 cents, and the net profit per acre $24.86. Not only were the gains in yield due mainly to 'lengthening the time of growth by preventing foliage destruction by late blight,' but the sprayed potatoes being more mature, were of better cooking quality." (U. S. Dept. of Agriculture, Bulletin 251, page 9—1906).

Stable and barnyard manure furnish heat as well as plant food; to get the maximum advantage of both it must be carefully treated and not allowed to waste itself on the desert air. Manufactured fertilizers can be used to great advantage if used wisely, but before paying forty dollars a ton, one must know exactly what the ton is composed of, what to do with it, and what the results are to be, for learning all of which there are abundant facilities in the books. The beginner is safe on stable manure, but on very dangerous ground with these commercial fertilizers. It is wise for him to use manufactured fertilizer only to raise green crops to plow under, in order so to improve the land as not to need to use them again. The bill for them is sure to come in, but a crop from them is not so sure to come out, although on early tomatoes, beets and cabbage and for high-priced crops they sometimes pay tenfold their cost.

Early vegetables pay best, and properly mixed manure heats the soil sufficiently to force them. But many gardeners now find it cheaper and more convenient to use hot water pipes, permanent or portable, to heat the soil.

Irrigation is giving astounding results, not the least important of which is that it is cutting down the size of the farms from the conventional 160 acres to a maximum of about 40 acres, running down to five acres, besides aiding and forcing plants and improving quality. That is a study by itself, but we can avail ourselves of its principles in the use of the farm-house waste and any pond or stream we find. It is as stupid to waste water, and especially dirty water, as it is to waste manure.

Ordinary farmers do not think it profitable to irrigate. A man who has push and the ability to handle a fine crop to advantage, finds it very profitable.

B. F. Calvert, of Willows, California, reports $1200.00 from an acre irrigated with his iron gasoline engine: the highest, watermelons, giving $300.00, down through strawberries, blackberries, Logan berries, tomatoes, to the lowest, cabbage, $100.00. The fuel costs only fifty cents a day.

The simple method of surface irrigation is to lay out at some distance—at least 100 feet—from the house a small sewage farm where the sewage may flow over the surface and slowly sink into the ground, which should have sufficient slope, and the soil should be porous, not retentive.

The liquid sewage, including kitchen and chamber slops, is conducted to this field in a watertight drain and then allowed to flow into shallow trenches. To avoid the overloading of the soil with sewage at any one place, the main distributing trench should be so arranged that it, and the irrigation trenches branching from it, may be temporarily blocked at any point to divert the sewage into one or more different trenches every day. In winter the warmth of the sewage will keep it in motion and the filtration will go on although the field may be covered with snow and ice.

The water from the roof may be used to advantage along with the soap-suds and slops, instead of simply keeping dampness around the house.

The simplest way to utilize kitchen slops is to pour them upon plants about the house in summer, in winter upon the soil, each time in another spot, so as not to supersaturate the surface layers of soil in any one place. However, marketable fruits and vegetables should not be carelessly allowed to come in contact with fresh sewage, nor should the irrigation field be near the well.

The Waring system, best adapted for climates where there is little frost, is to conduct the sewage underground in pipes with open joints and with trenches: but this is somewhat costly.

Here was an ingenious scheme for getting others to do his work—described by a writer in *Maxwell's Talisman*:—

"I put my new well on the higher side of the garden, with a large tub as a reservoir to get sun-warmed. The pump is near a road, and many people passed by daily. So I placed two cups and a notice on a box seat, which read, 'If you want a nice, cool drink, please pump a little while.' We had many a laugh over the notice at the well. Troops of ladies and gentlemen passing would stop and read the notice, and then start pumping. It takes over 200 lifts of the handle to fill the tub, but it has been filled five, six and seven times in a day, and during the warm weather the pumping was all done for us.

"The overflow of the tub went into the garden, and it saved me the expense of a windmill. The pump was well patronized, except on wet days. The bulk of this overflow went to the peas, beans, rose trees, black currant and some 500 raspberry and blackberry bushes."

Plenty of manure and then thorough cultivation make an almost complete protection against ordinary droughts. It is clear that when the "dry farmer" raises bumper crops on twelve inches annual rainfall, drought should have little terror for those who have forty to fifty inches per year. Always provided that we have no more land than we can treat thoroughly.

If the soil is cultivated carefully and intensively, it will hold and store water underneath the growing crop. Finely pulverizing and packing the seed-bed makes it retain most of the moisture that falls, just as a tumbler filled with fine sponge or bird-shot will retain much more water than one filled with buckshot.

The air sucks up the moisture from the earth unless we prevent it by a soil blanket or "mulch"; this also readily absorbs the dew and the showers. Water moves in the soil by capillary attraction as in a lampwick; the more densely the soil is saturated the more easily it moves upward, as oil "climbs up" a wet wick faster than a dry one. Put some powdered sugar on a lump of cut sugar and put the cut sugar in water; the powdered sugar will remain dry even when the lump is so wet that it crumbles to pieces; this shows how a mulch helps to check evaporation.

Grain and forage crops acclimated to dry conditions are being brought here from all parts of the world and are making lands productive which were formerly valueless.*

Much will be done with other crops also by dry farming; that is by plowing the soil very deep and cultivating six or eight times a season, thus reducing evaporation to a minimum and retaining all the moisture for the crops.

Thousands of acres in Montana grow good crops without irrigation. In Fergus County, for instance, the once incredible yield of 45 bushels of wheat per acre is grown without irrigation. Heavy crops of grain and vegetables are grown in the vicinity of Great Falls by dry farming.

In the strictly arid regions there are many millions of acres, now considered worthless for agriculture, which are as certain to be settled in small farms as were the lands of Illinois.

What irrigation has done for arid regions drainage will do for swampy, overflowed lands. According to the geological survey, there are

* Macaroni wheat will yield fifteen bushels to the acre with ten inches of rainfall. This is two bushels more than the average wheat yield in the United States.

60,000,000 acres of this sort of land now waiting reclamation. If half these swamps were drained it would increase the land values of the country by $300,000,000 and the crop values by more than $900,000,000; it would provide ten-acre farms for 3,000,000 families, thus putting about 15,000,000 people on lands now practically worthless. There is no doubt that the federal government will some day reclaim these lands as it already has the desert, so there is no real fear of the pressure of farm population in this country.

The gardener, like manufacturers and merchants, must devise new ways of packing and selling. One of the latest is the "family basket," devised by the Fullertons.

A crate holding six three-quart baskets was selected. The three baskets in the bottom contained beets, newly dug potatoes (the kind you can eat boiled in the skin) and cabbage. A partition over these and the top three contained peas, lettuce and cucumbers in one box, young carrots and onions in the third box. ("The Lure of the Land," page 88.)

This is attractive, but you will have to find a customer who will pay a fancy price for fancy goods, direct to you. It is apparently trivial things that do so much to increase profits: Eastern gardeners and farmers have much to learn yet about packing.

CHAPTER VII

MONEY AND TIME REQUIRED

The Teacher of Fools—Success—Failure—"Think-box" Secrets—Large Capital Not Needed—Specialties as Money-makers—The Value of Money—Equipment—Outlay and Income—Five-acre Investment—Ten Acres Costly—Cost of Starting—Union Wages in Different Cities—Hand vs. Horse Cultivation—Crops Every Month—The Condition of the Farmer.

MOST persons suppose that if their legs or their hands are not going, they are idling. But after all, as Mr. Carnegie says, "The man that does the work, the hewer of wood and the drawer of water, never gets rich." If working could make one rich, every mule would be a Croesus. Take time to consider what you should do and how you should do it, and read Agricultural papers, Department year books, and especially the farmer's bulletins. In our appendix is a selected list of those particularly adapted for the Intensive Agriculturist.

A LITTLE LAND AND A LIVING

The charge of the prophet against Israel was not merely that the nation had sinned, but "Israel doth not know, my people do not consider." If the printing press had been invented in those days, he would have added, "neither will they read good books."

Fools sneer at book-farming, but you need the best books by practical men. "Experience is a dear teacher, but a fool will learn of no other." A wise man learns by the experience of others.

Some people can't help being successful with plants, but while they have this gift they may lack the ability to teach you how to do it. If you watch them closely you will see that they love their work; that they handle their plants as tenderly as you do your children, and study them as closely as children should be studied.

Study the new methods of scientific, practical men, not the methods of ignorant yokels, who are still using the rack-rented hand-to-mouth methods that their grandfathers learned in Poland or in Ireland. You will find that if you use your powers of observation you can do most men's work better than they have been trained to do it.

Did you ever notice how little people know of their own business? Hardly a clerk can do up a decent package, or even tie the string so that it won't slip. So don't be overawed by experience nor by success. Even the successful man might have done much better if he had used others' brains as well as his own. The trouble is that men do not use their brains. As Dan Beard says, "It hurts the head to think,—try it and see."

To learn the rotation of crops and what things can be double cropped or companion cropped, what plants exhaust the soil and what restore it, and what weeds can be turned over for manure instead of being pulled up, and how deep to plow, is more work and counts for more than grubbing up stumps with a pick-axe.

You can spend indefinite labor, or money either, on an acre; your "think-box" must tell you how and where your labor and money can be most profitably spent and how to arrange so that the labor won't all need to be done at once. If you can't find that out, you will be like the "plain farmer," overworked and underpaid.

It does not usually pay at present, however, to work an acre to its full capacity; it hardly ever pays if you have to hire your labor, except for the merely mechanical part such as plowing. If there is plenty of cheap labor at hand you can exploit that; if not, you had better exploit only the land.

Large capital is not needed. Near every large city you will find market gardeners worth ten to fifty thousand dollars who started with little or no money. The uneducated Slavs, Hungarians and Wallachians are taking up the eastern farms long since abandoned to the mortgagees, and though they are not the most available lands that intelligence can get, the farmers are getting rich by cultivating them intensively.

A market garden is merely a big kitchen garden to supply the public as well as your family. For success or for profit, land near markets and transportation, manure, hot beds, crates, wagons and tools are needed. You can raise vegetables in variety and quantity sufficient to justify giving all your time to it.

So farmers, tired of small profits and hard work, should get away from staple crops to specialties. They will need only a few additional tools and fertilizers. A cash capital of $20 to $25 an acre would be enough for them, where a novice would need $80 to $100 an acre. A beginner in South Jersey on a five acre truck patch would need $500 to start and run till his sales of truck started. In Florida $95 an acre, in Texas $45, Illinois $70, around Norfolk, Virginia, $75 to $125, and in Long Island $75 to $150 are needed for tools, seeds, fertilizers and appliances, including rent of land, but not including labor.

Borrow the capital even if you pay high rates. It is a common superstition that money is worth only four or five per cent., and that all over that is a premium on risk. But the banks that pay that rate manage to make much more on what they get.

If you are making a thousand dollars out of

an acre that costs $500 you can certainly afford to pay much more than $25 a year interest on its cost, for the money will bring you far larger return than that.

You can work with only an axe, a plow and a spade, although a scythe and a sickle extra are handy, but to get the best results you need the best tools. Special work calls for special tools. They will pay for themselves many times a year.

Aside from axe, saw, plane, hammer, etc.; for market gardening, the following will be needed: 1 team of horses, $200 (though these may be hired); wheel hoe, $6.00—a wheel hoe saves backache—walking plow, $10.00; disc or cutaway harrow, $25.00; farm wagon, $50.00; cultivator (2 horse), $25.00; cultivator (1 horse), $8.00; wheelbarrow, $4.00; shovels, pick, matlock or grubbing hoe, $10.00; work harness for two horses, $25.00; spade and fork each $1.00; push hoe, 65 cents; watering can, 60 cents; rake and common hoe, $1.00; bulb-sprayer, 25 cents; trowel, 10 cents. A few good tools carefully selected and cared for are better than a number carelessly bought and then neglected. With proper care all these things, even the team, should be good for twenty years.

A broomstick or a tool handle sharpened, will make a dibble. Garden lines, to insure straight rows; labels; tomato supports, plant protectors and stakes can be home made, and thus lessen expense, but the following additional "boughten" tools will be useful:—

Wheel-hoe with seeder, $8.50; roller, $8.00; sprayer, $3.75; crowbar, $1.50; weeder, 35 cents. You push a weeder backward and forward and it cuts both ways. It's best on soft ground; on hard, use a push hoe. A horse hoe will save time on many crops; horse labor can be hired.

In Northern Long Island, if you do your own work, the cash outlay on an acre which produces 250 to 400 bushels of potatoes, usually selling for fifty to seventy-five cents per bushel, or from $125 to $300 an acre, is

Seed Potatoes	$10.00
Commercial Fertilizer	13.00
Spraying Against Blight, etc.	4.00
Total	$27.00

Hand labor, if you grow but one acre, might cost you $40, with wages at $1.35 to $1.50 per day. If you ship by rail to a consignee the selling charges would be forty to fifty dollars, leaving about thirty net profit. Where you do the work yourself, the labor cost, of course, goes to you as wages or additional profit. By thorough cultivation and care you can probably get 600 bushels without any greater cash outlay.

The cost of preparing an acre garden and the seeds for planting may be estimated as follows:

Counting Five Dollars per day for man and team, and two acres as a day's work, plowing	$2.50
One harrowing, at rate of 10 acres a day	.50
Manuring, 5 loads at $1.00 each	5.00
Seeds, one planting, mixed produce	4.00
Total	$12.00

The cost of seed is not a big item, but if you have to buy plants, the expense will be much greater. Mrs. Helena R. Ely in "Another Hardy Garden" says that $10 or $12 will buy all the seeds (excepting potatoes) required for a vegetable garden large enough to supply a family of eight to ten persons.

Two men could handle an acre with occasional extra help. On new rich land it will take two to three years to get fairly established. Worn out land takes longer. Asparagus and rhubarb take two years and bush fruits three, to become profitable. So you should lease for not less than ten years, or better, buy.

Give one acre to vegetables, one to small fruits, and one for buildings, poultry, cow and horse lot, etc. An active man should clear a thousand dollars a year, besides a good living, and be absolutely independent, unless he is located where some pirate can steal his profits.

When the writer asserted some years ago that he had known some "amateur farmers" to make not farm wages, which are about eighteen dollars per month and board (U. S. Department of Agriculture), but Trades Union wages, four dollars for every day of work, the judicious smiled, and the injudicious, including most of those who have failed in farming themselves, and even some agricultural editors, were tempted to scoff.

They said: "If that be so, how is it that the regular farmers can hardly make even one end meet, let alone two?" The answer seems to be first, that most farmers are really speculating in land; that few farmers know much about farming; and, finally and principally, that the farmer is pushed out by the high speculative prices of land to places where he ought not to be, and being, cannot succeed.

It might well be that some of the cultivators themselves were too enthusiastic about the results of their labors, but we will find that other facts cannot be explained if the returns were less than those stated. There is a story of an old Irish woman who was getting her mistress to write home for her. She dictated: "Oi get mate for me dinner onct a wake." "Why, Bridget," said the lady, "you get it every day." "Ah, yer leddyship, sure they'd never belave that at all, at all." That is like our four dollars—it was all of the truth that the public knew enough to believe.

T. B. Galloway, Chief of the Bureau of Plant Industry at Washington, calculates that where a man can sell his own crop of miscellaneous vegetables grown on five acres, his investment will be five acres of land at $250, $1250; greenhouse, 20 x 100, $1200; hotbed, sash and miscellaneous equipment, $500. He figures that the gross income from such a plant should be from $2000 to $3000 a year, or $1500 to $2000 net.

W. W. Rawson estimates that exclusive of the land and buildings it would cost $11,000 to equip ten acres for the highest intensive market gardening and require ten men in winter and twenty in summer, including three greenhouses 200 x 300. Therefore, don't invest in ten acres unless you have ample capital.

Where labor is scarce it is important to know how much time is needed for the thorough cultivation of each acre.

The Boston Report of the Industrial Aid Society (October, 1895, p. 5, Supplement), gives the average time spent on one-third of an acre as "eight or nine full days, some not spending over six days from the time of planting to harvesting inclusive." Call it nine, that is twenty-seven days per acre.

The estimate of Mr. J. W. Kjelgaard, New York, is that with the first plowing done, without the assistance of hotbeds, thorough cultivation of an acre of good land, one-half of which is in potatoes, will employ a good worker for twenty-four days a year.

Hon Seth Fenner, of East Aurora, New York, writes: "On a third of an acre of ground, plowing and fertilizing being furnished, half being potatoes and having no hotbeds, from six to eight days should be ample for a good worker to perform all the needed labor from inception to finish."

Mr. Joseph Morwitz writes: "Saturday afternoons and Sundays during the six months of work, or 39 days, ought to be just enough for one acre, if our test of 1901 on three acres Joint Account Patch were made the basis of calculation. These three acres cost us 1120 hours or 112 days of 10 hours, or 37 days per acre. Two crops per season. They were plowed, and I believe, cultivated, by the teams of the Society. The gardeners only planted and harvested and helped out the cultivating to some extent."

John G. Thompson, the fertilizer expert of Passaic, New Jersey, writes: "You wish an estimate of the time taken digging and planting. It will take an ordinary steady workman about four days to dig over and three days to plant a ¼ acre lot (28 days to an acre). This means persistent effort and continuity of purpose. Many persons will spend from two to five times as long and in fact never get through. They lack patience or perseverance, or both."

Mr. George T. Powell, the expert of the New York Committee on the State of Agriculture, says: "I think it will require thirty-seven days to cultivate an acre, one-half potatoes, and the balance in vegetables of an ordinary variety, including the gathering of crops. If the crops could be partly cultivated by horse power, the days would be reduced four or five in number. More labor is required on poor land without fertilizers."

Mr. Daniel B. Safford, an experienced gentleman farmer of White Plains, New York, figures it at thirty-five days for an acre.

We are safe, then, in counting thirty-six days as the time needed for an ordinary able farmer to cultivate a full acre plot.

A LITTLE LAND AND A LIVING

In Long Island City, New York, 56 acres divided into 84 plots averaging two-thirds of an acre each, required altogether 2016 days' work or 24 days for each plot. The total product was worth $4900.00, which gives an average wage of $2.43 per day. In Brooklyn, where the Gardens were as nearly a failure as anywhere, the average was $1.52. In St. Louis, Boston, Buffalo, Detroit, and Omaha the returns ran from $5.55 per day to 83 cents, or an average in the five cities of $2.70, and these places all started late and suffered greatly from drought.*

You can thoroughly cultivate an acre in 140 hours if you have a horse, or 250 hours by hand, say 14 or 25 days. At the South seeds can be planted every month; North from the first of March to August. Simple but complete tables are given in "Three Acres and Liberty" to have crops to sell every month. You can plant quick maturing crops and get almost immediate returns.

* The average product per acre was as follows: New York (commercial fertilizer used), $87.50; Boston (commercal fertilizer used on only 14 acres), $130; Brooklyn, $55; Buffalo (very poor soil), $48; Detroit, $60; St. Louis claims $200.

CHAPTER VIII

GROWING UNDER GLASS

Early Vegetables—A Sample Hot-bed—Heating by Fire—
Cost and Returns—Flowers Better Than Vegetables—
Greenhouses—Success.

TO get early vegetables you need a hotbed. For it you want the best soil and the sunniest spot possible.

All hotbeds are right-angled boxes covered with movable glazed frames and heated. The bed may be of any size or shape but the standard is six feet wide, since the stock glass frames are usually six feet long by three wide. The cheapest plan is to get some old planks, broken brickbat or stone, and piece together a box-like affair in proper shape, then lay on the sashes; the front should be at least ten inches above the ground and the rear fourteen inches, for drainage. Make it face south or southwest and protect it on the north. This will do to start with, but will last only two or three seasons. Cement walls extending to the bottom of the manure are best. Bank them with earth or straw to keep out the cold, and have mats or shutters for extra cold weather. The best material to heat the bed and the most easily obtained is fresh horse manure in which there is a quantity of straw and litter. This will give a slow, moist heat and will not burn out before the plants mature. Get all the manure you need at one time. Pile it in a dry place and let it ferment; every few days, work the pile over thoroughly with a dung fork; sometimes two turnings of the manure are enough, but it is better to let it stand and heat three or four times.

The soil should be equal parts of garden loam and well-rotted barnyard manure; tramp well the first layer of three inches; to make it entirely safe for the seeds add another layer of the same depth. Use no water with garden loam and manure if you can help it.

Before sowing the seeds, put a thermometer three inches deep in the soil of the bed. If it runs over 80 degrees Fahrenheit, do not sow. If below 55 degrees it is too cold; you will have to fork it over and add more manure. If the bed gets to hot, you can ventilate it by making holes with a sharp stick.

Another way to make a hotbed is with fire. On a large scale this is cheaper than manure. Start six feet from the east end of your hotbed and dig two trenches to about four feet west of the hotbed. Give them a slight taper to create a draught, and arch with vitrified tile, laying two bricks on each side, lengthwise, a little beveling, and one brick crosswise to complete the arch, and cover with dirt that was dug out of the trenches; to these the flue from a small stove supplies the heat. Unless you can get waste exhaust steam, steam heat is only economical in large houses. The care and expense do not pay except where the business is on a large scale.

There are ninety-five million square feet of glass in the United States devoted to vegetables, of which more than thirteen million feet are in New York State. Under favorable conditions, glass devoted to this work will earn an average of fifty cents per year per square foot.

The cost per sash is at least four dollars, or about sixteen dollars per frame, made up as follows: A well-mortised sash frame (4 sashes), $4 to $5; sashes unglazed, $1 each; glazing, 75 cents per sash; mats and shutters from 50 cents to $1 per sash, depending upon the material. These prices vary greatly, however.

An estimate for a one-acre garden to grow a general line of vegetables where half the acre is to be set with plants from hotbeds, is as follows: One-eighth acre of early cauliflower and cabbage, about 2000 plants, would require two 6 x 12 frames of four sashes each, allowing nearly 250 plants to each sash. These frames may be used again with 450 tomato plants, for the same area, about 55 plants to the sash.

Another frame would be needed at the same time, say for egg-plants and peppers, two sashes of each, growing fifty transplanted plants under each sash.

Two frames will be needed for cucumbers, melons, and early squashes; for extra early lettuce count sixty or seventy heads to a sash. Celery and late cabbages are to be started in seed-beds in the open. If spinach is grown in frames, the sash used for one of the late crops of endive, escarole, celeriac, and other October plantings, may be used through the following winter.

This makes five frames altogether, the cost from one to five dollars, according to make and material; twenty sashes and covers at say $2.75, $55; manure at market price, counting at least three to four loads per frame. This is a liberal estimate of space, and allows for all ordinary loss of plants, and for discarding poor ones. Most, or all of the plants, are to be transplanted once or more in the frames. Many gardeners have less glass.

Flowers pay better under glass than vegetables. A plant of carnations brings much more than a head of lettuce and suffers less from the competition of southern crops. But vegetables can be raised in houses that are too poor for flowers. Lettuce and tomatoes are the principal crops in amount and in profits. The greenhouse is also used for forcing plants which are afterwards transplanted to the open air. This develops them before they could grow outdoors, so that they are very early on the market, thereby realizing the highest prices.

That a small heated greenhouse is far better, and in the long run cheaper, than the manure heated hotbed, is the conclusion of many practical gardeners. The novice, however, will need to go slow in its use until he learns by experience. Greenhouses, like hotbeds, are expensive, varying according to size and materials used, but the increased returns from them justify their cost.

Small greenhouses, say 12 x 9 x 8 feet, with double walls, double-thick glass, plant tables, etc., are built in sections ready for putting together. They may be bought for from $80.00 to $115,000, and will more than repay the outlay, as the income is from 25 to 50 cents for every square foot of bench room, the prices, compared to those for open air products, being as five to one.

Nearness to market is most important. In large cities the chief fertilizer, manure, can be had for the hauling. The short haul is important, and above all, the gardener who is near the market can take advantage of high prices. If near enough to make two or three trips a day when prices are high, so much the better.

CHAPTER IX

ANIMALS FOR PROFIT

Animals on the Farm—A Snail Park—Frogs—Turtles—Bass—Pheasants — Dogs — Cats — Silver Foxes — Expenses and Receipts—The Busy Bee.

EVERY farm has its quota of animals for profit or for pets. Of recent years even pets are made profitable; there is a good living in rearing animals, preferably uncommon ones. This has opened opportunities to those who have land and yet are not able to carry on farming or even gardening.

Although the snail industry is practically unknown in this country, it has been in vogue since the days of Caesar, and flourishes now in Burgundy, France. Any plot of damp, limy soil can be made into a snail park by enclosing it with smooth, tar-coated boards to prevent the snails crawling out. The boards must penetrate the ground for a depth of eight inches at least, and have a ledge or shelf at the ground level, so that the snails may not burrow under them.

Strong stakes are driven outside the boards to prevent the wind blowing them over. A plot of ground 100 x 200 feet will serve for about ten thousand snails. The soil is plowed deeply in the Spring; the snails are then put in and covered with several inches of moss or straw, which is kept damp. As they eat only at night, their food,—lettuce, cabbage, vine leaves, or grass,—must be supplied daily about sunset. To improve the flavor of snails, mint, parsley, and other aromatic herbs are planted in their enclosure. The snail lays from fifty to sixty eggs in a year. Her nest is a smooth hole in the ground, where the eggs hatch in less than 20 days. The game is marketable when six or eight weeks old. They are ready for picking in October after they have sealed themselves up in their shells; they are then put on trays and kept in storehouse for several months, when they are brushed and cooked in salted water. They are shipped to market at once in wooden boxes holding from 50 to 200 each, and bring high prices as an epicurean delicacy.

Frog culture is successful only when the pond is large enough to be partitioned, thus separating the young from the old. The frog is a cannibal and will eat tadpoles and even the larva. The chief difficulty in raising frogs is to find the proper food for them, unless insects are abundant. There are many frog ranches near San Francisco and some do a thriving business, but as a general rule, commercial success has not attended the enterprise in this country.

Young diamond-backed turtles are cheap, while the full grown are enormously expensive, and the demand for them is constantly increasing. It is possible that purchasing young ones and maturing them would open a new line of profit. Common box tortoise and snapping turtles are raised for canning, and often take the place of the more expensive diamond-back.

Although fish culture is little known in this country except under Government auspices, yet carp, black bass, and trout are raised where conditions are favorable. It is comparatively easy to cultivate carp, although the industry has never attained such proportions here as in Europe. A pond with a mud or loam soil, and water of the same depth all the year, is all that is needed. In stocking a pond three females are allowed for two males. They spawn in the Spring, laying a large number of eggs, but only 800 to 1000 to each spawner prove fertile. Carp live to a great age and often weigh from 30 to 40 pounds.

Black bass give good returns. The small mouthed variety require a pond six feet deep in the middle and not less than two feet at the edge, with a sandy bottom and many water plants; 100 x 100 feet is a good sized breeding pond. As the bass needs a barrier behind which to spawn, artificial rectangular shaped nest frames are provided, two adjoining sides being 16 inches high, the other two four inches high. To keep the water in healthful condition the pond must be fed by a flowing brook protected from freshet disturbances. Bass feed upon minnows and in their absence must be supplied with fresh liver cut in threads like angle worms. Even then they must have minnows from September until they go into Winter quarters. Raising young bass or fingerlings to stock rivers and ponds is a profitable enterprise.

Trout are more difficult to raise, as cold running water and unremitting care are necessary

Although with pheasants, like all birds of the turkey family, the more ground they have to range in the less are they liable to disease, yet with proper care they can be raised on the home acre. We find authenticated cases where as many as sixty pheasants were kept in a house 10 x 50 feet with five yards averaging 10 x 25.

The chief difficulty in raising them is securing their food, such as flies, maggots, and ant-eggs. They also require green food like lettuce, cabbage, turnip tops, etc. The pheasantry should be placed on high, well-drained land with a southern exposure, and must be thoroughly protected from cats, dogs, and other small animals. Pheasants bring fancy prices and the supply does not begin to keep pace with the demand.

So much has been written about the raising of chickens that little remains to be said.

We do not propose to hatch out another article on the exhausted chicken subject. It is necessary to take with large allowance the articles in the poultry journals, as well as the Ananias stories on this and other subjects in certain highly illustrated "farm" journals.

A LITTLE LAND AND A LIVING

It is easy to figure out a profit in anything—on paper—and this is the way chickens figure out on paper:

Lot	Date				Sold C	Kept C	H
A	Jan. 1, '07	1 cock, 20 hens, will on Jan. 22, 1907 hatch 100 cocks, 100 hens. To preserve the Mormon proportion—sell 95 cocks. Begin with.......				1	20
						HATCHED	
		Hatch, etc.	(B)	Jan. 22, '07	95	5	100
A	July, '07	"	"	(C) July 22, '07	95	5	100
A	Jan., '08	"	"	(D) Jan. 22, '08	95	5	100
A	July, '08	"	"	(E) July 22, '08	95	5	100
B	Oct., '07	"	"	(F) Oct. 22, '07	475	25	500
B	Apr., '08	"	"	(G) Apr. 22, '08	475	25	500
B	Oct., '08	"	"	(H) Oct. 22, '08	475	25	500
C	Mch., '08	"	"	(I) Mch. 22, '08	475	25	500
C	Sept., '08	"	"	(J) Sept. 22, '08	475	25	500
D	Oct., '08	"	"	(K) Oct. 22, '08	475	25	500
E	not matur.	"	"	(L) July 22, '08	2375	125	2500
F	July, '08						
G	not matur.						
H	not matur.						
I	Dec., '08	"	"	(M) Dec., '08..	2375	125	2500
K	not matur.						
L	not matur.						
M	not matur.						
J	not matur.					7980	421	8420

Raising fancy chickens looks attractive, but it requires much experience to get the goods and much reputation to sell them; the risks are also great, so that unless one loves to fuss with fancy poultry it is not an encouraging field.

But to raise chickens of the ordinary variety, good for laying and for table use, much less fussing is necessary. A 200-egg incubator is filled twice in a season with fresh eggs, gathered from neighboring farms and henneries. Eggs laid the very day of collection are preferred, for those a week old are quite apt to be infertile. Practically all of those eggs will hatch and produce vigorous birds which will begin to lay early. No effort is made to secure any particular variety and only the most active and healthy are reserved for winter laying. Over-breeding has produced so many disabilities in fowls as in other animals that wise egg producers are now experimenting with cross-breeds. So far the plan has proven most successful.

Even a few chickens can be made to pay on a farm, particularly where the food is raised for them, so that nothing extra has to be bought;

the meat scraps from the table, the corn, cabbage, and other green things, wheat, oats and whatever food is used, will not be felt as an additional expense and the returns are therefore clear profit. A flock of 25 fowls, if properly housed and fed, will give eggs and chickens enough to add materially to the farmer's income.

There is more money in eggs than there is in chickens. Eggs out of season, like everything else out of season, bring high prices, and it is possible to nourish and even deceive the chicken into laying at unseasonable times. I do not know how good that is for the morals of the chickens, but it is good for our finances.

Intelligence alone will bring profits, and unless intelligence be reinforced by a careful study of the many valuable reference books and authorities, it will not secure large returns.

There is more profit in ducks than in hens, now that we know that they can be raised without swimming places. Ducks that are wholly land-raised have fewer feathers and more flesh, less oil and a finer flavor, and the demand for ducks as food has increased proportionately.

White pekins are the most popular breed because of their size and white meat; and moreover, they are splendid layers. They lay from 100 to 165 eggs a season and the ducklings are easy to raise.

Geese are another source of profit and like ducks can be successfully raised without water. The feathers of both are valuable, although goose feathers bring the higher price. In the autumn, pure white goose feathers, dry and in good condition, are worth about 60 cents a pound; gray goose and white duck, 40 cents each; gray duck 32 cents. Scalded stock brings from 3 to 5 cents a pound less. The small feathers are shipped in burlap or cotton sacks and, to avoid mildew, should be perfectly dry when packed.

Squab-raising is a lucrative business, but the returns are often exaggerated. Although pigeons naturally breed ten or eleven times a year, every egg will not hatch a big squab for the market. The inexperienced beginner, if he start with well mated pure Homer stock, may, with good management and close attention, clear

$2.00 a year on each pair of birds; but that is not likely.

Attention is turning to the "poor man's cow" as goats are called. The U. S. Experiment Station in Connecticut imported a herd of 50 Malta goats, which is the best breed of milch goats, for experiment. The goat is almost immune from tuberculosis, which makes its milk most wholesome. It is easily and cheaply fed, ten goats costing no more to keep than one cow. A good milker will give from 2 to 3 quarts a day and a child can give all the care needed. The does cost from $5 to $8 each when full grown; a young one from $1 to $3.

A further profit can be made from goats' hair, which is extensively used in the manufacture of so-called "camel's hair" goods. The best breed of goats for this purpose is the Angora, whose long, silky hair is much in demand for making the best grades of plush, silk dress goods, thread, etc. From three to five pounds of soft, silky hair is sheared from each goat annually, and the market price ranges from $2.50 to $7.00 a pound. Angora goats are not good eating, but they are splendid milkers; both butter and cheese are made from their milk. They are good breeders too, and cost but little to keep, although they need a herder and require shelter at night. The average price for a pair of Angora goats is $7.50; the male $5.00 and the female $2.50. A man who took up government irrigated land in Sirocco County, California, put nearly all his surplus cash, $100, into Angora goats, and soon owned all his land, built a house and started a bank account which is steadily increasing.

Swiss goats also are money makers. George L. Cook, of Winnetka, Ill., had a buck and sixteen does selected for him in the Saanen valley and brought to this country. It has proved a paying investment. Each doe gives from four to six quarts of milk a day, and the cost of keeping is very small. It is possible to build up a good trade in goats' milk in big cities, as it resembles mother's milk more closely than cow's milk does, and is better for babies. In Switzerland peasants drive the goats to the doors of their customers before milking them.

Of course goats are a venture, but they are

probably a better venture than cattle, especially if you get into the business before everyone else is in the race.

Dog breeding is not a new occupation, but many persons make money by it. It does not require much land or capital and there is always a profitable market, particularly for fancy lap-dogs, like the King Charles spaniel. To make profit on the breeding of retrievers or any sort of hunting dogs, kennels should be located in a game country and the puppies trained by a practical hunter.

Breeding cats is a newer and better paying venture than dogs. The purchasers are women who will pay any price for a cat that strikes their fancy, and as fashions in cats change frequently, there is always a demand. Parti-colored cats are just now in vogue, the favorites being Persian, tortoise-shell, and coon, but any oddity like the tailless Manx cat, or the lynx cats will find a ready market.

At Dover, Maine, there are two fox ranches where from twenty to forty silver foxes are raised each year on less than an acre of land.

They are not expensive to breed, as their food is chiefly sour milk and cornmeal or flour made into a small loaf, with a little meat once a week. They are clean animals and with careful attention are free from disease. They breed as well in captivity as in their wild state, six or seven being the average litter, and any man who has made a success of raising hens can succeed with foxes, which require no more space or care and are worth twenty times as much. Fine silver fox furs are worth $150 a pelt. A cold climate is needed.

A fox ranch should have a No. 16 galvanized wire fence, ten feet high with an overhang of 18 inches to keep the foxes from escaping. Stakes must be driven close to the fence to prevent burrowing. This and the purchase of stock comprise the whole expense for starting in a business whose gross receipts may mean from $3000 to $6000 a year on a score or two of animals.

But an easier business that is fast becoming popular is bee-keeping. More than the average living is made in raising honey bees, by careful and intelligent study. Any farming, gardening, fruit raising, or wild brush land will suit bees. Indeed, in Cincinnati and in Philadelphia, there are large colonies of bees kept on the roofs of houses in the busiest parts of those cities.

It is possible to begin with one hive, adding to it as circumstances permit. Even in a city one hive will yield 50 pounds of honey in a season, while under more favorable conditions from two to four times as much can be had.

The work should be taken up systematically and the would-be apiarist should take the best bee journals.

Bee-keeping has proved a very profitable venture to Mr. Stoughton Cooley of Maywood, Illinois. He started with four hives which yielded 400 lbs. of honey the first season, and at the end of four years they had increased so that his honey yield was 2700 pounds.

Mr. Cooley writes: "I worked at bee-keeping faithfully and thoughtfully and was particularly fortunate in small winter losses—one of the handicaps of the business. On the other hand, the foul brood, the worst of all evils to the bee-keeper, broke out among my bees and practically wiped out one season's profits; but I succeeded in stamping it out. I am well satisfied from my experience that I could attend to a hundred colonies (and I am lame), clearing from $500 to $1000 a season. After all, it comes to this, success in any line is for the few who will apply themselves faithfully."

Dr. C. C. Miller of Marengo, Ill., who is regarded as the American authority on bees, has harvested 18,150 lbs. of honey from 124 colonies. However, his warning is that the man whose sole aim is the accumulation of wealth would better let bees alone; but all bee-keepers agree that health, pleasure and a comfortable living are found in intelligently raising honeybees.

Profit in animal raising, like profit in all other lines of business, depends largely upon the time and attention given to it and the willingness of the worker to study, to observe and to profit by experience.

CHAPTER X

FRUIT GROWING

American Supremacy — Development — Improvement — Apples, quality, thinning—Peaches—Chances of Success — Protecting Grapes — Pears — Plums — Quinces — Cherries — Persimmons — Small Fruit — Extra Culture — Strawberries — Bush Fruit — Exceptional Returns.

FRUIT growing is rapidly increasing. The general climate of the United States is more favorable than that of Europe. Much American fruit is now exported because our fruit is more uniform and better than the European. Large quantities of apples go to England and Germany and we may look forward to an increasing trade in pears and other fruits.

In the last few years plantations of all sorts, orchards, gardens and nurseries, have increased wonderfully, from the extreme north to the south. But this does not necessarily imply improvement in quality. In the best fruit growing counties of New York more than three-fourths of the fruit plantations consist of very ordinary apple orchards. Few high grade pears, plums, cherries, apricots, grapes, or bush fruit have yet been produced. Cold weather in the spring tends to make conditions harder in New York than in the middle sections.

There is no profit in apple growing under average conditions and ordinary management, yet apples are the most popular orcharding fruit.

It costs only about ten per cent. more, mostly labor, to raise good fruit than poor, but it costs fifty per cent. more brains and brings a hundred per cent. more profit.

Prof. Samuel Maynard says we must make the trees grow vigorously, whether upon poor or good soil, as the first requirement. To get the finest fruit, we need a strong, deep, moist soil; good grass land well underdrained is best. An elevation with a northern or western exposure is better than a southern or eastern one. On land that is free from stones and not too steep, thorough and frequent cultivation will give the quickest and largest returns. On such land hoed garden or farm crops may be profitable while the trees are small, but after five or six years it will generally be found best to cultivate it entirely for the trees. Organic matter in the form of stable manure or clover-crops must be applied in the Fall or very early in the Spring to keep up the supply of humus in the soil. Good fruit crops may be raised on land too stony for other agricultural uses, if the soil about the young trees be well worked and the moisture retained by means of mulch. Grafting is so successfully done now that fruit trees may be made profitable early instead of waiting six or seven years for maturity. Windfalls, no matter how hard, can be made into sauce, jellies or pies, and thus furnish a return even before the fruit is ripe.

The new intensive methods are now applied to fruit growing. One of the most satisfactory of these methods is thinning.

In the Massachusetts Hatch Station experiment in thinning, Gravenstein and Tetofsky apple trees were chosen, only one of each being thinned and one of each left unthinned as a check. In the Gravensteins the yield on the thinned and unthinned trees respectively, was, first quality fruit, 9 bushels and 2½ bushels; second quality, 1 bushel, 2½ bushels; windfalls, 9½ bushels, 10½ bushels. In the Tetofskys the thinned trees gave 1 bushel windfalls, the unthinned 3 bushels; each tree gave a half bushel of second quality fruit, while of first quality the thinned tree yielded 2 bushels, and the unthinned tree none at all. First quality fruit brings 60 cents per bushel, second quality only 25 cents, so that the thinning more than doubled the value of the Gravenstein yield and increased that of the Tetofsky eleven times.*

Similar results were obtained with plums and in other places with peaches and pears and even with currants.

The advantages of thinning are that it increases the size and color, and gives a better flavor; it also increases the quantity of No. 1 fruit, lessens the windfalls, etc., and sometimes in-

* Such benefits did not show in every case, probably because of inexperience.

creases the total yield. It tends to check injurious insects and to prevent the spread of disease. Systematic thinning overcomes the tendency of fruit trees to alternate "apple years" and poor crops. It helps to establish a medium yield which averages more than the alternate abundance and scarcity both in quantity and in profit to the grower.

The peach crop is next in importance to the apple crop, and location is the chief consideration in planting an orchard. The orchards giving the largest returns are in Connecticut, Delaware, New Jersey, Maryland, Virginia and Georgia. For large crops it is well to choose land near large bodies of water as the temperature of the water prevents too early budding, and also delays killing frosts. While sandy, porous soil is best, peaches may be successfully raised in clay soil if plenty of humus be provided.

Some peach orchards are very profitable. Mr. Cornelius P. Swain, of Bridgeville, Del., says he picked 1100 baskets of Elberta peaches from 208 trees growing on about two acres, and sold this fruit for cash at the Bridgeville Station for

$1140. His orchard has had excellent tillage, is now eight years old, contains 500 trees, and has borne five crops, all of which have been profitable. The first paid $500, the second $850, the third $1000, the fourth $1200, and the fifth $1500.

Mr. Swain's orchard has had the best of attention, peas have been grown in it for the canning factory and cow peas and crimson clover for turning under, insuring good cultivation. The crops other than peaches have paid all the expenses of growing and tilling the orchard since it was planted.

Willis T. Mann, of Niagara Co., N. Y., says there is no branch of fruit growing in that section of country that is so much of a gamble as peach growing. The uncertainties of production, the very perishable character of the fruit and possible "gluts" in the market all tend to make the income uncertain, and it may vary from nothing to one thousand dollars per acre, after paying transportation charges and commissions.

Mr. Mann has sold a crop at five cents per pound and realized $900 per acre. A friend of his having a thirty acre orchard realized a few years ago $18,000 for the crop, $600 an acre. These, however, are possible, rather than probable results. Half these figures would still be considered high, and $250 or $300 per acre would generally be above rather than below the average. There has been a great increase in peach raising during recent years, and the fruit is now produced in large quantities where it was formerly thought impossible. During the past decade the number of bearing trees in the United States increased from 53,000,000 to more than 99,000,000, and the increase continues. The result is that our markets are often congested, and it is difficult to get satisfactory prices. Still, if the fruit raised is all choice, it will prove a very profitable end of the business. Land suitable for peach growing can be bought for $75 to $100 per acre in New York, and $25 more should procure the trees and plant them, thus making the initial cost $100 to $125 an acre. Five years of care, two or three of which should be self-supporting by the production of other crops on the land, would bring the orchard into bearing condition, and two or three years more would bring it to its highest value. It is difficult to state what the value of the orchard really is, because it depends more upon the man than upon any other factor. Two or three years of neglect will ruin the best orchard. Certainly the real value of the orchard would be greatly in excess of the cost of production.

At the Massachusetts and Colorado State Experiment Stations a new method of protecting peach trees has been successfully tried. This consists of laying the trees down in the Fall as soon as they have shed their leaves and the wood is well ripened. The earth is removed from around each tree in a circle about four feet in diameter, and the hollows are saturated with water and the trees are worked back and forth until loosened. They are then bent in the direction that offers least resistance until they lie on the ground.

The ground should be allowed to dry enough to handle easily before any further work is done; then the hole is filled in, the limbs of the trees are tied together and covered with burlap held in place with earth. A light layer of earth is then spread over the tree, affording ample protection to the tender buds.

Care must be taken that the covering is not so warm as to force the buds prematurely. In some comparative tests it was found that while in unprotected trees about 50 per cent. of buds were killed, only 10 per cent. of the buds of protected trees were destroyed. When they begin to open, the covering is loosened to admit light and air, but should not be removed except by slow degrees.

The trees are raised in the same way that they were laid down, softening the soil with water. In Colorado the trees are raised about the middle of May; in the East the season must determine the raising. Trees so treated will not stand unsupported, but are usually propped up at an angle, with two props to keep the wind from swaying them.

When this treatment is begun while the tree is young and persisted in each year, there is little or no injury to the root system, but it is not wise to try it on old trees. Indeed many practical farmers say that very few growers could use this method successfully even on young trees, and that it would, generally speaking, be unwise to attempt it.

Professor Waugh of the Massachusetts Experiment Station has done some important work on dwarf fruit trees; the "Miniature Fruit Garden" gives increasing promise, not only for its better resistance to frost and later bud opening, but for the beauty and convenience of these cut-back pigmy trees.

Fruits are second only to flowers in beauty and variety. We have miniature orange trees for decoration, why not baby apples and cherries also? Surely, too, a bouquet of fruit would afford as much scope for taste as one of "weeds." Poetry clings about the tree: if the poet clings, the profit might do it too. Think of the romance as well as the money in grapes.

No farm is complete without a grape-vine, though only fruit enough for the family be grown. The vine can be planted near the house so as not to interfere with crops, and at little expense an arbor can be built. This will provide a cool, shady retreat in hot weather. Grapes form a delightful table fruit and can be made into delicious and lucrative jams and jellies for winter use; there is little grape jelly now on the market.

At Fredonia, Chautauqua County, New York, in "the Grape Belt" of this State, there is a colony of wine growing Italians holding title to 1758 acres of land and conducting a thriving business. There are four hundred families or about 2000 persons, and they have increased the value of the land in that section from $35 to $150 per acre. A large proportion of them own their own farms and are not therefore paying tribute to landlords. Frugality, thrift, and persistent, faithful attention have made their vineyards profitable to an extent unknown elsewhere in this country. The fresh grapes bring from $20 to $30 a ton, the wine from a ton of grapes fetches from $30 to $36. The cost of production runs from $8 to $10 per acre, and an acre produces three tons of grapes annually after the first three years, so that the grower has an average profit of $75 per acre.

These farms and vineyards have done much to

relieve the congested Italian quarters in Buffalo and other interior cities, and may well prove a satisfactory solution of the Italian problem of New York city, when further extended. The Italian will only work in a gang. He is too civilized to go out by himself among strangers. But when settled in groups with their families, Italians succeed famously as farmers.

The plum, the pear and the quince may be made profitable under conditions like those for peaches, and recently cherries have come into prominence, although a more precarious crop than any of the others. Care must be exercised in choosing a site for the cherry orchard, high ground and free circulation of air being necessary. Because of poor packing, Eastern cherry growers have suffered from competition with California growers, who have reduced packing to a science, although Eastern cherries have a better flavor.

The persimmon is another fruit not generally grown; in fact the native variety was little known until the Japanese persimmon was introduced. Some of the native wild varieties are of exceptionally fine quality, and careful selection and grafting can make it the equal, if not the superior, of its Japanese relative.

A persimmon will ripen wherever peaches thrive well, although they do best in Southern States. They are usually picked and sold before fully ripened, and few consumers know just how delicious the ripe fruit is. A market for them could be created in any large city. Like fresh figs, the damage in transit makes them scarce in the Eastern market. Let someone learn how to pack them. Figs preserved in glass bring excellent prices.

In the "Small Fruit Culturist" Andrew S. Fuller says, "The cultivating of small fruit as a distinct feature in horticulture commenced less than twenty-five years ago. We may well feel proud of the progress we have made in small fruit culture, but the limits have not been reached, and for those who may wish to enter this field, there is many an unsolved problem to work out.

"With a constantly increasing demand, and no apparent prospect of our market being fully supplied, many have turned their attention to the cultivation of small fruit. It offers as wide and safe a field as any other branch of business except natural monopolies. In many instances with an annual expenditure of twenty-five dollars per acre, a return of only one hundred dollars is obtainable, while upon the same soil and with the same variety, if fifty dollars had been expended, the return would have been three or four hundred dollars. All experiments show that *extra culture* is far more profitable than what is generally termed *good culture.* Many fruit growers, for the purpose of extending their business, increase the number of acres, when, if they would double the depth of those which they already possess, they would obtain the same increase in product without the expense of more land, and the extra travel and travail of cultivating two acres, when one acre might produce the same results.

About four out of five growers in northern New Jersey prefer a sandy soil for strawberries, and in the southern section, where the greatest number of growers are located, fully two-thirds preferred sand loam. Yet from statistics gathered in the southern section, the State Experiment Station showed that the average returns per acre from clay loam were much heavier than from sandy soil; a given number of growers averaging 3223 quarts per acre from the clay against 2829 from the sand. For large yields, clay loam is most satisfactory, but for early crops a sandy loam with southern exposure is best.

Prof. Fred W. Card in "Bush Fruit" says that "land which will yield profitable returns in bush fruits can be found almost everywhere, if the soil be not wet and heavy. Red raspberries and blackberries succeed well on comparatively light soils, provided they retain moisture, while dewberries thrive on very light sand, and currants and gooseberries are at home even in heavy clay. Stable manure is the best fertilizer. Wood ashes, cotton-seed hull ashes and muriate of potash form a useful supplement.

In answer to an inquiry for an average yield of blackberries per acre, fifty growers in different parts of the country reported from 1280 to 10,000 quarts, the average being 3158 quarts or more than ninety bushels per acre.

Instances of admirable yields during one season are common enough. One grower in a small town in central New York sold $500 worth of berries from half an acre, and M. A. Thayer, of Spark, Wis., raised 200 bushels per acre and made a profit of about $250.00.

Of course, one great consideration in fruit growing is to be near a good market. Shipping long distances often causes loss by decay, while the freight charges materially lessen the profits.

To grow fruit for family use and have some for sale would not require much land. One of the U. S. Agricultural Department bulletins gives a plan for a small fruit garden, about 60 x 80 feet, on which the following could be grown: Six peach trees; six cherries; six dwarf apple trees; six plums; twenty blackberries; forty black caps; forty red raspberries; three hundred strawberries; thirty-two grape vines, planted at intervals of ten feet all around the plot; and eighteen dwarf pear trees. If properly cared for, such a garden would yield a large return.

CHAPTER XI

HORTICULTURE

The Market — Violets—Greenhouses—Diseases—Roses—Chrysanthemums—Poppies—Flowers in the Street—Sweet Peas and Wild Flowers—Orchids—Plants for Renting—Floriculture.

IN every city there is a growing demand for flowers. Roses, violets, carnations, and chrysanthemums are now most popular. But to make horticulture profitable you must in this, as in everything else, study your market before deciding what to raise. If you give all your thoughts and energies to raising one flower you can produce better results than by raising a variety. If you raise enough you can sell direct instead of through a commission man and so get better prices. There are always good markets somewhere, and flowers are shipped from New York to Chicago, Buffalo, Boston, Philadelphia, Baltimore, and Washington, and vice versa.

Intelligent effort is the road to success, very

little capital being needed. Few among the leaders in floriculture started with more than $500, and many of them with less. A dozen men who have been in the employ of one New York florist, some of whom got only twenty dollars a month at first and afterwards started in a small way for themselves, are now making a substantial living.

A lover of flowers can succeed in this business better than in any other with as little capital. In the last ten years the business has doubled, and while many have gone into it, the profit they are making shows that supply has not equaled the demand and that is not likely to be overdone soon.

An acre of soil under glass pays fifty times as much as an acre out-doors. There are eight to ten million square feet of glass in the United States devoted to carnations alone, and about seven million dollars worth of this one flower are sold each year.

Mrs. H. C. Reynolds and Miss Nina F. Howard started a violet farm at Glencoe, Illinois, in 1905. Their friends predicted failure, but they studied the best methods of growing, got the best possible soil and a greenhouse. They have made it a big enterprise. You can do as well.

Mrs. Reynolds says: "I am only a beginner. The first year I had a small house of 3000 plants and was so successful that I have begun now on a larger scale. I have 210 x 165 feet of land on which I hope some day to have six houses 25 x 150 feet each. I have now two, of 5000 plants each. Of course the market is here, as all our double violets came from New York and however beautiful they were, they had lost their perfume on a long journey."

W. J. Harrison, a druggist at Lakewood, N. J., has a violet frame on the ground in front of his show window; it gets enough radiated heat from the cellar to give a winter crop of violets with almost no attention, and violets sell remarkably high in Lakewood. You may make a start in some such way and grow up to greenhouses.

For women who like country life and wish to work at home, violet growing offers very great inducements. The work is easy, but constant attention and great care are necessary.

The temperature of the violet greenhouse must be kept between forty-five and fifty degrees. It must have a system of ventilation so arranged that it can be operated from within or without, as fumigation with the deadly hydrocyanic gas is sometimes necessary, for insects. Violets need all the sunshine you can give them in December and January and as little as possible at other times. They are subject to four dangerous diseases, all difficult to exterminate when once started, and known as spot disease, root rot, wet rot, and yellowing. The best preventive is to get strong vigorous cuttings, to give careful attention to watering, cultivation and ventilation, and to destroy dead and dying leaves and all runners as soon as they appear.

Chrysanthemums are in great demand, particularly the large sized, oddly colored ones. They are extremely decorative and last so well that their continued popularity is assured. The importance of the rose and chrysanthemum business is well indicated by the annual shows.

Poppies are fast coming into favor as a cut flower, although they wither quickly. But by picking off all blooms in the evening, cutting the new flowers early in the morning and plunging them immediately into deep water, they are made to keep fairly well.

Mr. Powell, of Fairhope, Alabama, the single tax colony, has shipped some gladiolus "spears" to New York in the winter, when they readily bring ten cents each wholesale, though they sell in the Spring at five. He says that fine bulbs will sell better still.

In flowers, as in all garden products, it is the thing out of season that pays best. In summer when gardens are all a-bloom, there is a smaller demand for greenhouse flowers, but Fleischmann of Fifth Avenue quotes the following winter prices for cut flowers in New York:

Chrysanthemums, $2.00 to $5.00 per dozen; American Beauty Roses, $1.50 to $5.00 per dozen; Violets, $1.00 per hundred; Carnations, Killarney Roses, Brides and Maids, Richmonds, $1.00 per dozen; Lilies of the Valley, $1.25 per bunch of 25.

Flowers that are sold cheaply in the streets are the discarded stock of swell florists. You will find that they are either fading or revived with salt and will not keep. That they are so peddled, shows that everyone wants flowers, those angels without souls. Even common flowers will bring good profits and are easily grown. We have only to supply a want to find our place in life.

Sweet peas, a favorite flower with many, may be grown out of doors in the summer where the soil is of good depth and quality. Mayblossoms, autumn leaves on the branch, and even goldenrod are brought into town and sold at good prices.

Bachelor buttons, cosmos, and even nasturtiums, which you can't keep from growing, if you just stick the seed in the ground, or lilies of the valley, hardly to be got rid of when once started, all find a market, if fresh. And the best of it is that one does not even need a greenhouse to grow them and many other dear familiar garden blossoms.

No soil, however hard or apparently barren, is too poor for flowers. Eben Rexford, in "Four Seasons in a Garden," says that the ground of a city yard which he, with the aid of two boys, made into a garden in one season, was so hard that they had to use axes to break up the clods. It was then allowed to stand two days for the action of sun and wind, was fertilized by stable sweepings, and planted with Petunias, Phlox, Calliopsis, Nasturtiums, Zinnias, Asters, Poppies, Marigolds, Sweet William, and Morning Glories. It was a source of delight the whole season through. No home acre should be without its flower garden, as a joy and a revenue raiser.

Some florists have ventured into orchid raising, but the business has many drawbacks. To make any profit, the flowers must be shipped in large quantities and with great regularity, else customers are lost and expenses eat up returns. To get such a supply of orchids would require a large investment and involve much labor.

By keeping a supply of ferns, palms, and rubber plants constantly on hand and renting them out for social functions, weddings and other occasions, many florists make considerable money.

Near the larger cities a thriving business is

done in tree planting, which is everywhere on the increase. Many florists now raise young trees and plants for sale to people improving their grounds, planting orchards, etc. This nursery business, as it is called, is a special department of horticulture and bears the same relation to the commercial florist or orchardist that seed growing does to the market gardener. The largest nurseries in the country were started on small capital, but soil and climate enter so largely into success that the business is not over-crowded.

The ever increasing demand for bulbs has led to a great extension of the business of raising them, but America has a comparatively small share in it. Holland is the greatest producer, large areas being devoted exclusively to the raising of lilies of various sorts. Bulbs require deep, rich, warm, and highly manured soil and the most careful attention. The high prices paid for land in Europe has led to the most modern and ingenious methods of increasing soil-production. The same necessity has not hitherto existed in this country, but it exists now, and it has become necessary to get the greatest possible returns, by intensive farming and gardening, from the smallest possible piece of land.

Flower growing is well adapted to women because, generally speaking, they have a light touch and a capacity for details. The work is light, has few disagreeable features and the rewards are sure. As the aesthetic qualities of our people develop, the raising of flowers for sale will be proportionately greater.

CHAPTER XII

BUILDING

Clearing the Land—Lumbermen Buy—Profitable Trees—Building a Home—"Hickory Bungalow"—Portable Dwellings—Remodeled Buildings—Comfortable Cabins—Water Supply.

IF there is but little timber on the land, a man can clear it at odd times, but even if covered with trees there is no cause for discouragement. Lumbermen will readily buy timber by the acre, cut it and haul it. Or the lumber may be used in building the house where there is no suitable shelter for the family on the "home acre."

Good judgment should be used in cutting trees, considering not only the market value, but also their qualities as shade trees or revenue producers. Elm and sycamore trees near a house are better than a stock of awnings and shades, and nut and fruit trees are better than income bonds. Sugar maples ten or fifteen feet high

A LITTLE LAND AND A LIVING

can be transplanted or sold. For the rest, black locust sticks can be sold for insulator pins; cedar sticks for hop and bean poles, "curly" or "bird's-eye" maple or birch are in great demand for furniture. Stumps may be burned or dynamited out. Dynamiting is the quicker method, but it needs an expert and is expensive.

The cheapest and simplest way of getting rid of large rocks is blasting by fire. A fire is built on the top of a rock and kept burning until the boulder has been thoroughly heated. When it cools it will break easily. If the rock is below the surface, a trench must be dug around it, the rock slightly raised by levers, and a fire built in the trench so that the heat may get under the rock. Some people who have burned rock out, advocate pouring water upon them while hot, to cool them quickly, while others again have found that the natural process of expansion and contraction is all that is necessary.

This method, which has been endorsed by miners, farmers, and others, will recall the ancient story that when Hannibal crossed the Alps he removed the rocks by heating them and pouring vinegar over them, thus causing them to break. The story has long been regarded as legendary, but recent events seem to give it an air of truth. So far as has been reported, nobody has actually tried Hannibal's method.

When the clearing has been made, we tackle the house building. Recent experiments show that this need not be the expensive undertaking it once was. Even if one be unwilling to live in a tent or put up a mere shack, there is no call to expend thousands of dollars on a house.

Comfort, convenience and utility are the important considerations, and these may be secured in a bungalow. In the June, 1907, issue of "The Village" Mr. William Jeffery says:

"To get rid of household care, live in an American bungalow. I know, for I have tried it. A bungalow is a low, compact, rural building, with small cellar and large porches, and ours is built alongside of a large hickory tree on account of which we selected that site for Hickory Bungalow. The tree is cooler than any awning in summer and the delicate tracery of its branches is a delight in winter.

"The main structure is 30 x 30 feet and faces south. The piazza takes eight feet in width across the front. The kitchen is at the end, leaving 22 x 30 feet living room; from the end of this a strip twelve feet wide is screened off by frames of pine and burlap secured by mouldings, for bedrooms. On top of the screens and reaching to the ceiling we have a nice grill which permits of sound ventilation and adds dignity to the interior. The floor of the large room upstairs is laid with two smooth faces on top of the beam; two chestnut posts support it. The room upstair is 13 x 30 and is reached by a detachable combination ladder and stairs. Some of the boys sleep up there. A window in each end provides ample ventilation and light.*

"We built a large chimney with an open fireplace, but we found that the open fireplace was a delusion. It looked pretty, but did not give enough heat; so we installed a wood stove with air chambers to improve radiation; when the weather is far below zero we can run the tem-

---

* Mr. Jeffery has now added a separate bungalow for his boys; it is a great scheme—but "that is another story."

perature up to 85 degrees indoors. We inclosed the piazza for the winter with a detachable frame containing storm door and windows. This space is handy for the children to play in in extreme rainy weather.

"All this cost less than $1500 and gives more majesty than a regular house that cost $3500 to build. It has already saved us nearly the whole of its cost.

"We have not seen our family doctor since we came here. My family consists of six boys and a girl. One of the boys, six years old, had never been out of the leech's care until we came to the bungalow. And now the doctor sends a message that if all his patients take to bungalow life he will have to go into some other profession.

"This sounds like a testimonial for some patent medicine, but there is no patent for fresh air and life on the ground floor, so I give it a free 'ad.' We can leave our windows and doors all open even in smart cool weather and get more air in a minute than we used to get in our house in a month. No corridors or halls to obstruct ventilation or to keep clean, no parlor or reception hall, one-story kitchen 12 x 14, cellar underneath to keep coal and supplies, hot and cold water, range, sink, etc.

"I cut off eight feet of the piazza for a room. I sleep in it with open windows; temperature about the same as outdoors and this is the first winter in thirty-eight years that I have escaped quinsy and tonsilitis.

"Life in a bungalow means more than simple living. It means free and easy fresh air; family not too close together; (for our bungalow gives more space than a two-family house), one's own garden, poultry, dogs, and horse; fields, flowers, woodland, and stream for the children; the robin, blue jay, black bird, and thrush, for constant companions. Fun is expensive, but pleasure does not cost a cent.

"One of the chief advantages of getting into a new community is that one thus finds his opportunity for improving conditions, shaping better tax laws, securing good roads, up-to-date facilities, free schools, lectures, entertainments, etc., all of which add to the charms of country life. Our life in a bungalow has been a fine success. It is at Berkeley Heights, about 500 feet above sea level, in a beautiful country with woodland and river, yet only 26 miles from New York. There are plenty of such places.

"I am proud of my bungalow and will be pleased to show it to anyone, as a type of American house especially suited to the intensive farmer."

Except in lumber districts wooden houses are becoming more expensive annually and are being superseded by houses built of concrete blocks. These are fully as cheap now as frame or brick houses and can more easily be made cold and vermin proof.

Then there are the portable houses costing from $300 for four rooms, upward. These are desirable where land is taken on a short lease or for experiment, although they can be adapted for year round use at a moderate additional cost. These different types of dwellings show that the cost of building a home need not deter anyone really anxious to get back to the land for a living.

Of course if there are buildings on your acre, you may count yourself lucky. No matter how old or dilapidated the house, it can be made habitable by the expenditure of a few dollars, or a good barn may be made into a house at a low cost.

Whether one build or alter, one should study plans so as to get the most for one's investment. It is easy to spend a few hundred dollars in making an attractive house, but if it cost a great deal to heat, or the sun cannot get into the living rooms it will be expensive to maintain. A fireplace in the outer wall of a house will do little to warm the interior, no matter how big the fire in it; while, if the chimney is placed in the middle walls, half the fuel will keep the whole house comfortable. Get some woman's aid in arranging your house. The house is a woman's domain, generally speaking, and she can give valuable hints to the best architects.

Some advertising architects will guarantee their estimate of cost—no charge for the plans if the house costs more than they say; that will help to keep one out of a financial hole.

An important item is the water supply, if the place chosen is away from town water systems. If there is a good spring on the acre, pipes can be laid from it to the house and a pump in the kitchen sink will do the rest, but if you can get a fall of ten or twelve feet from spring to kitchen, no pump will be needed.

If you have to drive a well, the average cost is about one dollar per foot. A barrel or tank over the stove where the water can be pumped gives a good water supply to the kitchen.

Where the water system does not admit of a bathroom and toilet, the outdoor closet should be built in a modern way. The contents if mixed with earth or litter may be used as a garden fertilizer.

Make a shaft, say 2 x 4, down the side or at the corner of the house to form the toilet room. Sections of 12-inch vitrified pipe form a receptacle, the end with the collar being uppermost and touching the underside of the seat. The lower end is supported by bricks or stone a foot from the ground. Once a week put a wheelbarrow load of good earth, rotten sods, or leaf mould into the box, at the sides of the pipe, where it can be reached by raising the seat. Then take a barrow load out at the bottom and the waste is completely absorbed and broken down by the bacteria. The soil must be full of humus and friable. Sand or road dust will not do. If the grade permits, a longer shaft is better, as the earth will then take longer to pass through; with one length disintegration and deodorization is complete in ten days. The vitrified pipe is clean, non-absorbent and confines the mass to small limits. It can be washed with hot water. Two scoopfuls (less than a quart) thrown in when the closet is used, is sufficient. There will be no drafts in such a closet.

CHAPTER XIII

CO-OPERATION IN OPERATION

Toil Without Reward—"Back to the Land"—How to Get Land—Co-operation—Man's Natural Job—Organization—In Europe—In America—Fellowship Farm—The Arden Colony—Farming a Business—Changes Imminent.

HARD, anxious work, with little to show for it in the end, is the portion of the average man. Life is something that was thrust upon him unasked and must be maintained at any cost. It seldom occurs to him that there is anything either beautiful or wonderful in it. He begins to drudge in youth and for years the daily round of rising unrefreshed from sleep under conditions that make rest impossible, to spend hours in a workshop or factory and then return to his cramped, airless quarters, goes on without hope of change. How incomprehensible to him the joyous cry of Browning "Oh, the wild joys of living!" A man who said this to his fellow workman would likely be thought crazy. But it was that these wild joys of living might be more generally known that the "Back to the Land" movement was started. It has gathered force so that nothing can now stop it, and the only thing remaining is to show the eager seeker for land how he may get it and how to work it to best advantage. Nothing can give a man such a feeling of independence as owning the land on which and from which he lives, and there is no need of despair because he is unable to buy what he wants at once. There are ways of making small sums count. Co-operation is one of them.

Co-operation is wealth. Given certain resources, or raw material, human labor on the "together" plan produces manifold results which can never be realized by individual separate effort. Indeed, the key to our 20th century economic progress is combination, organization, co-operation—these three—and the greatest of these is co-operation.

The basic resource, or raw material, for the field of human co-operation is the land. Everyone instinctively knows that his natural "job" is on the land. Those who are engaged in other occupations than tilling the soil, as Emerson says, "are using a makeshift and are only temporarily excused from their real calling." Land and labor are wed, whosoever puts them asunder commits sacrilege; for in their union is health, wealth, and happiness—in their severance is disease, glut, and hunger, arrogance and misery.

Therefore, workers, get land.

How can the working class, poor men, get land? Every acre of available land is held by owners and costs more money than the average worker can accumulate; he cannot buy a square foot at a time, as he does a hod of coal or a bundle of wood. How can such a hand-to-mouth wage-worker buy land, even only an acre?

By combining his slender surplus with the little savings of others, and together pooling a monthly sum, hardly more than they spend for "beer and baccy," and organizing for the purpose, a group of from twenty to fifty working men can buy the best farm in their locality.

This is being done extensively in the German

Empire, especially in Prussia and Bavaria, as also in America, particularly in Southern California—the Dos Palos colony, the El Capitan colony, and the British colony, besides the Coburn Township in Texas, and New Clairvaux and Fellowship Farm in Massachusetts being successful examples. Most of these colonies organize simply to secure for individuals one to twenty acres, the colony making a collective first payment on the tract of land bought, then carrying the interest, taxes, and subsequent payments of principal until the titles are cleared, when they dissolve their organizations, each person or family securing its separate holding. But after disorganization, often the co-operative plan continues in their buying and selling of commodities and products.

The most recent and typical example of co-operative land buying and land using is on Fellowship Farm, in Westwood, Massachusetts. Here, in 1906, a group of forty socialists, gathered from different States and two from Scotland, organized (after they had raised $1000 to pay down) and secured a beautiful, $8000 seventy-five acre farm, 14 miles from Boston, via trolley and steam cars. Some put up as much as $100, and others could only advance a first payment of $2.50 toward the required $1000 down.

The $7000 on mortgage was arranged so that it can be discharged with all interest and taxes paid in ten years—each member of the group to pay as dues $2.50 per month until he or she has paid a total of $300,—or collectively $12,000. For this $300 each is to receive a warranty deed to a lot of about an acre on a fifty foot road, besides a fortieth interest in the collective holding of some twenty-five acres with buildings, and the benefits and profits of common enterprises. There is a provision for the *individual* free and clear of the group, and also for fellowship and corporate effort and life. Since the organization nearly an extra $1000 has been raised for improvements on the collective property, and besides a large cottage worth about $1800, several cottages costing from $200 to $300 have been built, showing that for $500 or $600, accumulated at the rate of $5 or $6 a month, a working man with a small family can secure land, and a house of his own on it, and a job from which he cannot be "fired"—can become moderately independent and free—by *co-operation*.

The motto of the Association should be:

> Get three acres and live by it.
> Get a spade and try it.
> And get out of debt.
> Get off the back of the workers.
> Get out of the power of the shirkers.
> Get up and Get.

Co-operation has made great strides in Great Britain, although colony life is only in its infancy. H. D. Lloyd in "Labor Copartnership" says: "The co-operative movement after fifty years of struggle, has had years of living prosperity and a greater prosperity is coming into view. It has achieved an economic footing of a hundred millions of property and enlisted at least one-seventh of the working-classes in its ranks. It is an established religion; for co-operation is not a method of business merely, but an ideal of conduct and a theory of human relations. Co-operation has won the right to be accounted the most important social movement of our times outside of politics. It is of course only a half truth, but the world needs half truths to make up its whole truth."

There are tracts near all good markets where communities could be established. The value of co-operation and combination has been learned partly through the trades unions and partly through the success of big corporations. As soon as men learn to look upon farming as a business to which business principles should be applied, co-operative farming will spread rapidly.

The earth can be made to yield many times more abundantly through the application of intensive methods; the same methods applied to effort will give still more abundant returns. This is proved by all the experiments in co-operative farming.

But the people must have room near to one another to make co-operation natural and easy.

There will be a revolution in our farming plans and in our farm life just as soon as the people wake up to the fact that the land about our towns and cities, nearly all of which is now held idle for speculation, is the land out of which they should get their living. Nine-tenths of Greater New York is not in use.

The normal and healthy growth of our communities is strangled by a system which hugely rewards men, at the expense of the workers and improvers, for keeping the available land vacant. Those lands about every centre of population in every civilized country are grossly undervalued for taxation, often at only five per cent. of what they would sell for.

Bay Shore, Long Island, for instance, but by no means the worst one, is choked by the Lawrence Estate of some seven hundred and fifty acres on the west; the bulk of which is worth $600 per acre, but assessed at $135,000, including improvements; by the South Side Club's some eight hundred acres on the east, which could not be bought at $500 per acre, but is assessed at only $15,000. And the village is cut in two by the Dominy lands of some eight acres, worth upwards of $10,000 per acre, and assessed at about $18,000 for the plot. Meanwhile, the houses of the working people are assessed up to about 80 per cent. of their real value.

As Shearman shows in his "Natural Taxation," taxing things that people make hits the farmer hard, though he does not know it. Little of his property can be concealed from the assessor, and he pays directly and indirectly much more than his due share of taxes. But that is as nothing compared with the indirect injury he suffers by being forced out by the high prices of suburban land to a distance from his customer and from the opportunities and advantages of city life. Isolation and the cost of hauling are bad enough. But far worse is the waste of energy and decreased production resulting from the holding out of use of these lots that are near the markets. The farmer is forced out to the wilds, and his customers are forced into the cities, and between the two is this desert of speculation.

A millionaire consumes but little more food stuff than a mechanic ought to consume and would consume if his opportunities of earning were not cut off by the holding of land out of use and the consequent crowding of the workers into slums and tenements. The farmers' natural customers, the great plain people, are not able to buy the things they need. This is mainly because their rents or the prices they have to pay for their home sites are so high, and because those lots where homes and shops and factories ought to be are kept idle, and mines, coal lands, clay pits, quarries, and sand banks, that are now needed to employ the people, are unworked, or only partly worked.

The suburbs and the country districts are undeveloped or half developed, because the money raised by taxation is insufficient to pay laborers who need the jobs, for making good roads, bridges, streets, water works, and other necessities of civilized life. So the growth of sections near the cities is checked till the speculators cut their own throats, and there is a fall in prices or "dull times in suburbans."

There is neither reason nor justice in allowing owners of valuable lands to hold them almost untaxed until the pressure of population forces the prices up to fancy figures which tempt them to let them go—fancy figures which must be paid, indirectly and directly too, by those who make things.*

We most stupidly and unfairly discriminate in our assessments in favor of those who do nothing but keep land idle, and against those who develop and improve it. When the natural opportunities and resources of the lands which lie out of the cities are assessed for taxation, as the law even now requires, up to their real and full value, we shall have done much to solve the problem of the congestion of towns and the unemployed, and we shall have, not for the four hundred but for the forty millions, an era of prosperity of which we have not yet dreamed.

* Land in Greater New York averages over $200,000 in value for each acre.

CHAPTER XIX

TO START SANITARIUM WORK

Outdoor Life Effective—Dr. Trudeau's Plan—Bad Conditions—Convalescents Work—The Earth for Men—Location Important—The Superintendent's Value—The Money Needed—Supplementary Industries—Preserving, Baking, Selling, etc.—Start Now.

LITTLE has yet been done in providing outdoor employment for tubercular patients, although the experience of the Vacant Lot Gardens in individual cases has shown it to be most effective.

Dr. Trudeau's sanitarium in the Adirondacks, however, has just initiated an Industrial and Gardening Association where patients will be able to employ themselves in the open air, eventually to support themselves and to learn a trade which will keep them outdoors.

To return an arrested case to the bench or desk is generally to re-pronounce the sentence of death from which the unfortunate has been reprieved. It is, therefore, to be hoped that there will be a large and rapid extension of this common sense principle.

The plan undertaken will provide occupation, training and support for those who are convalescent from consumption. Little work is to be expected from those who are ill, but when the patient is cured, or the disease arrested, the person requires outdoor life and the opportunity of self-support without returning to indoor labor.

This is no experiment; it has been done on a charitable basis in more than twenty places in the United States. That it can be done and how it is being done by the use of very small patches of land intensively used, is amply shown elsewhere. The paralytic, rheumatics, drunkards, defectives, cripples, women, even children, can shift for themselves, if they get to mother earth. Says Lyman Abbot: "For one man who can find a job there are thousands who can do a job well when it is found for them." We must find it for them, but how?

People will work and work effectively, when they have the opportunity and intelligent supervision, and when they see the results and get the whole proceeds for themselves. Get a piece of land; ten acres will do, and fifty is not too much; it should be near a town or an institution where the product may be sold.

It is better to pay five hundred dollars for an acre of land close to residences than a hundred dollars an acre for outlying land, because it is cheaper to get stable manure to it, and it is more accessible to the cultivator, besides being easier to show and closer to the market; and because the rise in the value of the land is greater and more certain.

A temporary shelter or house should be put up for superintendence and storage; tents and shacks will serve for the cultivators who desire to live on the land. But most of them will go to and fro from their residences.

Everything depends upon the superintendent, who should be capable of undertaking the whole affair; finding the land; raising the money; having the land plowed and getting stable or other manure put upon it. Then let the superintendent see all those who might help to buy seeds and plants or get them donated. Explain the plan to the cultivators and apportion plots to each, a quarter of an acre or less, up to even an acre, in accordance with the probable capacity of each.

The superintendent, he or she, should have hotbeds prepared in advance, mark out the plots and meet the cultivators by appointment at the ground. Instruct them as to the preparation of the soil, what they should plant and how; which way the rows should run, and what must be put in each; how deep the seeds should be planted, and every other detail of culture. The superintendent should be constantly on the ground and know and be patient with the peculiarities and shortcomings of the cultivators. There will be stupidities and obstinacy enough to try the patience of Job, but "the guides should not be angry with those who go astray." Any who cannot or will not work simply forfeit their plots for neglect or insubordination.

When all this is done, the enterprise is started. The money needed will be:—enough to make at least a partial payment on the land, to put up the shelter and a stable; to get a pair of horses and a harness for hauling manure; to buy seed, a farm wagon and some kind of carriage, a plow, harrow, weeder, sprayer and other tools, for all of which (exclusive of the land and buildings) five hundred dollars will be ample.

The superintendent's salary should not be less than eighteen hundred dollars a year, as on that officer's ability and intelligence the whole thing depends. Even if a greenhouse be not used (as it is not essential), and if there be no industrial features added, the superintendent will find ample work in the winter in making the report, lecturing, writing and otherwise extending the scheme.

Here is an opening for any capable man or woman anywhere—for there is unbounded room for such institutions and charitable people will give freely to them.

To employ those who cannot wait even the month or six weeks needed for radishes and such quick-growing crops to mature, there should be a co-operative plot on which they may be given work at such a rate per hour as their labor may be worth, the crop to belong to the Association.

For this purpose seven hundred dollars should be enough—say a thousand. Another fifteen hundred should be added to all this, to avoid running short. The building of a house is a later matter, but it will naturally follow. Even if but half a dozen should be found to take up individual plots to begin with, there would be no cause for discouragement, as the plan here as elsewhere will be justified by its works, and experience shows that as the first of the succession crops begins to come up, more applicants for the land are found than can be accommodated.

Some will be found who can take up detached bits of land, under the superintendent's direction and preparation, and take care of them on their own account. Many know how to raise chickens, flowers, fruits, or small animals, and can have a chance to do so. To care for an acre intelligently does not take over a month's work altogether during the entire season even of unskilled labor,* and when the rudiments of agriculture are learned, employment for spare time can be found in car-

* See Chapter VII.

ing for the yards and plots of town residents, and in selling the products.

Natural supplementary industries will grow up around the headquarters, as they ought to grow around any intelligent cultivator's plot, such as preserving fruit and fine vegetables, drying kitchen herbs, like thyme, marjoram and summer savory; cake baking, candy making, nut shelling, birch bark and rustic work. Those who can themselves sell direct at retail, early, fresh, well selected vegetables or flowers, will of course get the largest prices.

What is necessary is to make a beginning—now. To wait for another year is needlessly to sacrifice not only well-being, but lives.

Halifax, Nova Scotia, owned a large tract of land in connection with its city prison. It was wild, rough, rocky land supposed to be incapable of cultivation. No effort was made to grow anything upon it. In 1880 the city appointed a new governor, a man city bred, with no experience in farming, but he had faith in the possibilities of God's good earth and prepared to prove it. With the cheap labor at hand he began at once to clear the land. He used large quantities of dynamite to blow out rocks and stumps. Where, as in many places, the soil was too thin to support vegetation, he had earth hauled. He fertilized thoroughly with stable manure (got free for the hauling from livery stables), and also with a compost made from the contents of vaults mixed with earth sods and stable sweepings, and thoroughly rotted. Lettuce, onions, carrots, beans, peas, chives, beets, potatoes, turnips, parsnips, cucumbers, cabbage, squash and celery figured in the crops he raised, to say nothing of the fields of hay and forage raised for the horses and cattle.

Besides the saving to the city in cost of food stuffs, the male prisoners who worked in the sunshine throve so well that this, with the change in their diet made possible by the farm products, entirely banished scurvy and other diseases which had been prevalent hitherto. The plan of giving each prisoner a plot to cultivate for his own use has not been tried, partly because it had not occurred to the governor, and partly because the prisoners have such short terms that the one who planted would seldom reap. But the results have once more established the truth that no land is barren; a little earth, a quantity of fertilizer, considerable intelligence, a study of agricultural journals, good seeds and timely care, will secure a crop anywhere.

Philadelphia has recently carried the experiment of farming into hitherto untried fields; it has set some of its insane charges to work at cultivating the land, and the results have been most satisfactory. Many of the workers were hopelessly insane, but the benefit derived from healthful, outdoor work was most marked even in the worst cases, while many who suffered from milder forms of dementia were completely cured. The value of the products on the Byberry farm was fully $10,000, and larger returns are expected next year. The workers themselves were delighted with the change and took much pride in their work. Healthful exercise in fresh air led to a natural fatigue so that the insane workers readily fell asleep at night, and "Nature's kind restorer" helped to heal their troubled minds. This experiment will doubtless be repeated by other cities where the problem of properly caring for the demented is being carefully solved. If the excitement, speculation, fear and depression of modern city life induce insanity, as they undoubtedly do, it is reasonable to expect that a return to healthful, normal conditions of living—getting back to the land upon which man depends for everything—will do much to effect a cure.*

* There is a general impression that farmer's wives are especially subject to insanity. This may sometimes occur from loneliness and drudgery, but a somewhat superficial investigation of Massachusetts and Iowa statistics showed that they are less subject to insanity than factory hands.

CHAPTER XV

THE PROFESSION OF FARMING

Agriculture, Past and Future—The Average Farmer—The Boy and the Garden—The Trained Farmer—Salaries Await Him—Fresh Discoveries—Fields for Investigation—Grown from the Best—The Profit of the Earth—Monopoly Conditions—The Value of the Farm—How to Proceed—The Aim of This Book.

AGRICULTURE has been neglected as a profession because of the isolation of the farmer forced to low priced land at a distance from those with whom he must trade; the bright boys will not stay for such a life as that, so they go to the cities, leaving stagnation behind them.

The future of farming lies with our boys. Whether it shall become the chosen calling of specially trained young men with all the added dignity that such training could bring to it, or remain what it has hitherto been, concerns them even more than us.

Although tilling the soil is the oldest occupation, it has been the last to feel the necessity and benefit of special training. For this reason it has not taken its rightful position. But its possibilities have attracted specialists who have shown that no career is more inviting or more lucrative or more dignified than that of the skillful foster-father of plants.

It is a common saying that "the Southern farmer spends his life fighting grass, to raise cotton, to buy hay."

In just the same way the Northern farmer worries over "them pesky briers," blackberries and raspberries, that grow by his fences, and roots them out to plant corn that brings him fifteen dollars an acre profit: if he would encourage the briers he might reap five hundred dollars an acre from them.

Says the President of the Chicago Great Western R. R. Co.:

"The eccentric and witty Lorenzo Dow was in the corn-hog belt when he said:

"'The average Western farmer toils hard, often depriving himself of needed rest—to raise corn—for what? To feed hogs—for what? To get money with which to buy more land—for what? To raise more corn—for what? To feed more hogs—for what? To buy more land. And why does he want more land? Why, he wishes to raise more corn, to feed more hogs. And in this circle he moves until the Almighty stops his hoggish proceedings.'

"But in my judgment, working early and late to raise more corn, to feed more hogs, in order to buy more land, is not farming, but speculation. The great fault of American agriculture is too much land."

That is passing away and the farmer is learning that the small farm near the town is the money maker.

So far as farming is concerned, big-scale production is not increasing, but the opposite. For better transportation, better culivation, and above all, better education is changing farm life. Now we must make the land attractive to the boys so that they shall not go "back to the land," but will never leave it.

Children brought up in city tenements tend to become sickly and vicious, whereas on the farm they have a chance to grow into vigorous, self-respecting men and women. It is the city that breeds or attracts most of the pauperism and crime. The country has its own healthful life. To give children a fair start in life it is necessary that they should live close to "the good brown earth."

The school gardens, the home gardens and the individual plots belonging to the boys are the best means of influencing the children, and there has been a remarkable growth of these here and in Europe in the past seven years. The effect is moral as well as economic. Even in the worst neighborhoods there has been no stealing, and the children themselves have been watchful to prevent injury from either carelessness or dishonesty.

Through such agencies we shall discover what children have a natural bent for farming, and in due course parents will encourage that taste as they now encourage other tastes and inclinations. In the past the boy's experience of farming was doing hard work at the behest of another, and getting no direct benefit. Besides, farming was a blind thing even to the farmer, how much more to the uninstructed boy?

The one idea many people have with reference to the child and agriculture is—weeding! Weeding that, with the "total depravity of inanimate things," always requires to be done when other and larger interests are absorbing the child's mind.

Now, the fact is, agriculture is one of the most fascinating of occupations to children.

Hundreds of cities within the last few years have taken up the "School Gardens" idea, and they report this year thousands of applicants for garden plots they have no room for. In Watertown, Massachusetts, for example, a town of only 1200, they have 150 garden plots and 20 on the waiting lists, besides 100 "home gardens," i. e., garden plots in the children's home yards. These "home gardens" are properly a post-graduate course, the child needing the instruction received in the community gardens to fit him to work alone.

The crux of the matter is, if you want a child (or an adult) to take an interest in agriculture, give him entire charge of a plot of ground and let the product of his labor be his. The cultivation of the soil has been proved an effectual means of development along wholesome lines when the members of any institution are allowed to get the most they can out of the soil, and to have what they get. On the other hand, it has no more value than the "weeding" when they are treated simply as hired hands.

When a boy shows an inclination toward gardening he should be given a plot, shown how to plant it, encouraged to make a study of agriculture, and allowed all the returns. This will make him regard farming or gardening as a business, rather than merely as an occupation; that is the first step toward restoring agriculture to its proper place.

Take in good agricultural journals, secure reliable books and discuss possibilities with him. Children treated in this way surprise us by their original and practical ideas. A boy who has gathered the fruits of his toil, read and studied good agricultural papers, will want to know more than these can give him. He should then be given a few years at an agricultural college. When through he will know more of soils and their possibilities to start with, than his father knew at the end of a long and laborious life.

Having special training for his calling, it will take on added dignity in his eyes and eventually in the eyes of his neighbors. A "farmer" will not be synonymous with an illiterate boor, but may take his place with lawyers, doctors, ministers and others who are specially trained for their life work. Hitherto farmers and mothers have been the only persons with important work who have had no scientific training.

Farming under modern methods is as desirable a profession for a girl as for a boy, and any bent in that direction should be as thoroughly developed in the one as the other. It opens another healthful, natural business for a girl.

In New Jersey the Baron de Hirsch Trustees have a farm at Woodbine, and Rutger's College, New Brunswick, has courses in agriculture. Others are following suit.

There is little danger of that profession becoming overcrowded. Millions of acres of land in the country are yet unbroken, and millions more, under so-called cultivation, can be made to produce ten-fold by scientific methods.

Large land owners, Irrigation and Reclamation Companies, are inquiring for graduates of all such schools. Government experiment stations and the great railroad systems are on the lookout for them. At one time railroads were content to offer good lands to the farmer and let him make what he would; now they have learned the importance of helping him to see what could be made if he followed the best plans. Scientific men are awaking to the limitless field of science applied to the land. Every day brings us authentic reports of extraordinary and important discoveries. For example, Prof. Waugh finds that soaking seeds in beer promotes vigor and growth, which is interesting to our prohibition friends, though it does not appear that beer soaking improves men.

It has been found that the agaves, of which the century plant is one, furnish food, drink, soap, clothing, cordage, (sisal) paper, sticks, light, beams, medicine and ornament. They grow in Mexico, and there is no doubt that the increasing price would make it worth while to find a species that will grow in our deserts.

We know almost nothing of nature's resources; there are uses for many plants now regarded as pests, but no one has yet investigated; quantities of roots, fruits and herbs not now in use are valuable edibles. We call them weeds because we do not yet know their uses.

Someone might write a useful book and gain reputation and position by summarizing and popularizing for Americans the vast literature on Intensive Farming which exists in France, Germany, Italy, Denmark, Holland and other countries. The literature of farming is extensive. We have over 400 farm periodicals in the U. S. The Astor Library in New York has some ten thousand volumes on agriculture in which one can lose himself, if not his mind; but there is little of the lore of foreign countries accessible at present in this country; however, there is enough for newspaper and magazine articles, the reception of which might encourage the expenditure of time and money to get these most important and interesting facts, about which we know, at present, practically nothing.

No soil is worthless or really barren, if we bring the soil and the right plant together. Fertilization and irrigation will make any soil productive for almost anything.

The greatest rival to Holland in bulb growing is now the Puget Sound country. Mr. George Gibs established the business there after investigating soil and climate conditions. Despite failure at first, he persisted in his efforts and is now growing rich.

Those who do not know and will not learn, cannot hope to make a living off three acres or thirty acres. But he who studies, observes, experiments, finds it pays him well to cultivate even one acre thoroughly.

You are constantly urged to think and study, but you must do something also and do it now. Thinking without doing is dreaming. You don't want to be like Bobby:

"What was that terrific noise upstairs early this morning?" inquired Bobby's father as the boy appeared at the breakfast table.

"Well," explained Bobby, "I dremp I was a duck, an' when I woke up I had swum off the bed."

Seed improvements offers a big field. One practical man found that potato scab can be combated by simply exposing the seed tubers to that great germ destroyer sunlight, and that this hastens their growth.

A minister inherited a wornout farm of fifteen acres with a soil of reddish gravelly clay. After studying the best and most modern farming literature he started a dairy farm. He kept 30 head of cattle, raised his own "roughage," sold milk, bred cattle, and cleared more than $1000 a year per acre. The cows were kept in the barn the year round as his land, being near the city of Philadelphia, was too valuable to be used as pasturage.

The time is at hand when the principles we are laying down in this book for specialties will be applied to the great staples, and we shall be able to double and redouble our yields.. Even now it is true of corn when farmers give special attention to it. The instances given in Chapter

V of the results of the *Agriculturist* contest show what can be done. When corn is grown for the ears, and not for the stalk, farmers will plant only such seed as has proved most prolific, and the stalks will be cut back, that all the plant's strength may go into producing more ears.

If you happen to have poor land and you have brains enough, you can make your neighbors wish their land was as poor too, as one man, Mr. E. McIver Williamson, of Mont Clare, South Carolina, did. He took advantage of the fact that his land was very poor to stunt his corn, and put all the strength of the plant into the ears.

This was the way he went about it. He planted the corn in the poor subsoil and the best it could do was to grow to between two and three feet high, then when the time came for the ears to set he piled on a rich top soil, in other words, fed the plant all it could use, and all the strength of the plant went to the formation of ears; not as in ordinary cases, mostly to stalk and leaves. The consequence was he had a much larger yield per acre than the average, and his envious neighbors complained that they could not do that because their soil wasn't poor. Mr. Williamson has made a special study of corn growing. He says: "No farmer is so rich that he can afford to till poor land, nor any land, except in the best manner, planted in the crop best suited to it." In 1904 at an expense of $11 per acre for fertilizer, he averaged 84 bushels yield to the acre, the best single acre yield being 125 bushels.

Farmers are awaking to the value of selected and specialized seeds. One hundred and twenty-five students took the special examination for corn judging certificates at the Iowa State Agricultural College at Ames last year. In the corn growers' contest there were nearly eight hundred entries. The corn sale followed, during which the world's record was established.

The prize lots brought fine prices, sales of ten ears at twenty-five, twenty and fifteen dollars being common. The champion ear of corn last year, which was exhibited by H. J. Ross, was sold for eleven dollars. The grand champion ear grown this year by D. L. Pascal, was offered at auction. Starting at ten dollars, the bids were raised successively, five dollars at a time, until the price of one hundred and fifty dollars was reached, when the ear was secured by the former owner. It was the Reid Yellow Dent variety, weighing nineteen ounces. Each kernel is valued at thirteen cents, and the purchase price is at the rate of eight thousand, eight hundred and fifty dollars per bushel.

The champion ten ears of corn ten years ago sold for thirty dollars, which up to that time was the highest price paid for seen corn.—*Farming,* May, 1907.

It is poor economy that keeps the smallest and poorest potatoes for seed. The breeder of animals does not keep the weaklings for breeding purposes. He selects the best. Just so corn should be planted from seed secured from the stalks that produced most.

In 1899 six pounds of Swedish select oats were planted in Wisconsin. In 1905 there were harvested 9,000,000 bushels. In localities where it was unprofitable to raise ordinary wheat because of slight rainfall, macaroni wheat was planted. In 1905 the yield was 20,000,000 bushels.

If anyone thinks the profit of the earth will

come to the cultivator without intelligent and persistent effort, he is a fool. No owner of land, unless others require it to live upon, can make money by neglecting it.

There is no scarcity of land to feed the world even at the old wasteful rate, but it is not accessible under present conditions, being held by speculators. The cry that there are too many mouths to be fed by the world's supply is no more true than that there is more grown than could be used. It has been one of the anomalies of life that while thousands are starving in the cities, grain is being used for fuel by the farmers. It must ever be so while natural opportunities are in the hands of a few. It is a condition created by monopoly, a monopoly in no danger at present of indictment for restraint of trade.

But the capacity of the land now under cultivation has never been tested. That is why agriculture offers a field for the trained farmer today, unexcelled even by the possibilities of invention. All men cannot be inventors like Edison, but he who succeeds in multiplying and improving the ears of wheat which may be grown in a given area will be doing mankind a service that will not be reckoned lightly.

The new agriculture is only in its infancy and the demand for farmers who can do what was once thought impossible is growing daily. The farmer who is content to go along in the old rut, adding acre to acre and merely scratching the surface of each, is as sure to be left behind in the race as the merchant who prefers sailing vessels to steamers, and who declines to use the telephone.

That is why farming offers such an opportunity to the trained young man or woman. There are few plums in the law compared with the many aspirants; the supply of physicians is already in excess of the world's real needs; there are more ministers living on insufficient incomes than are enjoying comfortable salaries, the average being less than $450 per year; there are few paying professorship for the educationalists: and even where opportunities in these lines do exist, they are accompanied by serious drawbacks such as boards of government, or political intrigue. But the farmer is independent. People must be fed and the demand for the best food increases daily; there is no limit to the possibilities of his profession.

But what is the farmer to do who has remote, unsalable, perhaps exhausted land, and little capital? He thinks he cannot move nearer to the city; his home, neighbors and associations are all tied to the land *adscripti glebae*; what shall he do to get out of the rut and into better conditions?

If he is going behind year by year, he might better move before all is gone. If he is "just making a living," he can put the most of the farm into pasture or into hay, or let it run wild in the hope of a rise in land value, and specialize on his orchard, or on berries, or on a few acres of fine vegetables. He will not make any less than he makes now, and he will find some hope for himself and give some hope to his boys.

The most important thing to teach to-day is how to make the greatest profit from the least land. When the farmer has learned that, he will have no cause to fear the absorption of farms into large holdings. The value of the farm lies

A LITTLE LAND AND A LIVING

chiefly in the farmer, so that a very small tract by intensive cultivation will give a good living and provide for old age. Even two acres will do this and more than this.

Discoveries in other fields have revolutionized methods and created fortunes. Agriculture offers unlimited opportunities to the bright boy or girl who makes a business of scientific farming or gardening. The world is waiting for the agricultural discoverer.

With the help now freely given by the Government, the agricultural colleges and experiment stations, you can help the world to great advances.

It is necessary to know something about land and how to get the best returns from its cultivation before you take up farming as a life work. It is therefore wiser to keep your present position until you have studied and become familiar with the valuable and helpful literature on agriculture, and saved at least a small sum to meet emergencies.

Go slowly at first. If you have no practical experience, try first to get a garden where you can give it your spare time without losing your present position. Then as you get skill and experience you can give it your whole time and will not be helpless, if, for any reason, you lose your job.

To show what has been done and what can be done on small areas; that the life of the farmer, now so laborious and unprofitable, offers the widest opportunities for success under new methods, is the object of this book. If it help any to take up the pursuit of agriculture as a business, if it draw even a few from the crowded, unhealthful life of the city tenement, to the free, health-giving life of the fields and gardens, if it afford a ray of hope to the discouraged toiler, it will have accomplished its purpose and fully repaid its author.